THE GOLDEN RATIO

ALSO BY MARIO LIVIO

The Accelerating Universe: Infinite Expansion,
the Cosmological Constant, and the Beauty of the Cosmos

THE GOLDEN RATIO

The Story of Phi,
the World's Most Astonishing
Number

MARIO LIVIO

BROADWAY BOOKS
New York

Broadway Books titles may be purchased for business or promotional use or for
special sales. For information, please write to: Special Markets Department,
Random House, Inc., 1540 Broadway, New York, NY 10036.

PRINTED IN THE UNITED STATES OF AMERICA

BROADWAY BOOKS and its logo, a letter B bisected on the diagonal,
are trademarks of Broadway Books, a division of Random House, Inc.

Book design by Caroline Cunningham

ISBN 978-0-7679-0815-3

In memory of my father

Robin Livio

PREFACE

The Golden Ratio is a book about one number—a very special number. You will encounter this number, 1.61803 . . . , in lectures on art history, and it appears in lists of "favorite numbers" compiled by mathematicians. Equally striking is the fact that this number has been the subject of numerous experiments in psychology.

I became interested in the number known as the Golden Ratio fifteen years ago, as I was preparing a lecture on aesthetics in physics (yes, this is not an oxymoron), and I haven't been able to get it out of my head since then.

Many more colleagues, friends, and students than I would be able to mention, from a multitude of disciplines, have contributed directly and indirectly to this book. Here I would like to extend special thanks to Ives-Alain Bois, Mitch Feigenbaum, Hillel Gauchman, Ted Hill, Ron Lifschitz, Roger Penrose, Johanna Postma, Paul Steinhardt, Pat Thiel, Anne van der Helm, Divakar Viswanath, and Stephen Wolfram for invaluable information and extremely helpful discussions.

I am grateful to my colleagues Daniela Calzetti, Stefano Casertano, and Massimo Stiavelli for their help with translations from Latin and Italian; to Claus Leitherer and Hermine Landt for help with translations from German; and to Patrick Godon for his help with translations from French. Sarah Stevens-Rayburn, Elizabeth Fraser, and Nancy Hanks provided me with valuable bibliographical and linguistic support. I am particularly grateful to Sharon Toolan for her assistance with the preparation of the manuscript.

My sincere gratitude goes to my agent, Susan Rabiner, for her relentless encouragement before and during the writing of this book.

I am deeply indebted to my editor at Doubleday Broadway, Gerald Howard, for his careful reading of the manuscript and his insightful comments. I am also grateful to Rebecca Holland, Publishing Manager at Doubleday Broadway, for her unflagging assistance during the production of this book.

Finally, it is due only to the continuous inspiration and patient support provided by Sofie Livio that this book got written at all.

CONTENTS

1

PRELUDE TO A NUMBER

Numberless are the world's wonders.
—SOPHOCLES (495–405 B.C.)

The famous British physicist Lord Kelvin (William Thomson; 1824–1907), after whom the degrees in the absolute temperature scale are named, once said in a lecture: "When you cannot express it in numbers, your knowledge is of a meager and unsatisfactory kind." Kelvin was referring, of course, to the knowledge required for the advancement of science. But numbers and mathematics have the curious propensity of contributing even to the understanding of things that are, or at least appear to be, extremely remote from science. In Edgar Allan Poe's *The Mystery of Marie Rogêt*, the famous detective Auguste Dupin says: "We make chance a matter of absolute calculation. We subject the unlooked for and unimagined, to the mathematical formulae of the schools." At an even simpler level, consider the following problem you may have encountered when preparing for a party: You have a chocolate bar composed of twelve pieces; how many snaps will be required to separate all the pieces? The answer is actually much simpler than you might have thought, and it does not require almost any calculation. Every time you make a snap, you have one more piece than you had before. Therefore, if you need to end up with twelve pieces, you will have to snap eleven times. (Check it for yourself.) More generally, irrespective of the number of pieces the chocolate bar is composed of, the number of snaps is always one less than the number of pieces you need.

Even if you are not a chocolate lover yourself, you realize that this example demonstrates a simple mathematical rule that can be applied to many other circumstances. But in addition to mathematical properties, formulae, and rules (many of which we forget anyhow), there also exist a few special numbers that are so ubiquitous that they never cease to amaze us. The most famous of these is the number pi (π), which is the ratio of the circumference of any circle to its diameter. The value of pi, 3.14159 . . . , has fascinated many generations of mathematicians. Even though it was defined originally in geometry, pi appears very frequently and unexpectedly in the calculation of probabilities. A famous example is known as Buffon's Needle, after the French mathematician George-Louis Leclerc, Comte de Buffon (1707–1788), who posed and solved this probability problem in 1777. Leclerc asked: Suppose you have a large sheet of paper on the floor, ruled with parallel straight lines spaced by a fixed distance. A needle of length equal precisely to the spacing be-

Figure 1

tween the lines is thrown completely at random onto the paper. What is the probability that the needle will land in such a way that it will intersect one of the lines (e.g., as in Figure 1)? Surprisingly, the answer turns out to be the number $2/\pi$. Therefore, in principle, you could even evaluate π by repeating this experiment many times and observing in what fraction of the total number of throws you obtain an intersection. (There exist, however, less tedious ways to find the value of pi.) Pi has by now become such a household word that film director Darren Aronofsky was even inspired to make a 1998 intellectual thriller with that title.

Less known than pi is another number, phi (ϕ), which is in many respects even more fascinating. Suppose I ask you, for example: What do the delightful petal arrangement in a red rose, Salvador Dali's famous painting "Sacrament of the Last Supper," the magnificent spiral shells of mollusks, and the breeding of rabbits all have in common? Hard to believe, but these very disparate examples do have in common a certain number or geometrical proportion known since antiquity, a number that in the nineteenth century was given the honorifics "Golden Number," "Golden Ratio," and "Golden Section." A book published in Italy at the beginning of the sixteenth century went so far as to call this ratio the "Divine Proportion."

In everyday life, we use the word "proportion" either for the comparative relation between parts of things with respect to size or quantity or when we want to describe a harmonious relationship between different parts. In mathematics, the term "proportion" is used to describe an equality of the type: nine is to three as six is to two. As we shall see, the Golden Ratio provides us with an intriguing mingling of the two definitions in that, while defined mathematically, it is claimed to have pleasingly harmonious qualities.

The first clear definition of what has later become known as the Golden Ratio was given around 300 B.C. by the founder of geometry as a formalized deductive system, Euclid of Alexandria. We shall return to Euclid and his fantastic accomplishments in Chapter 4, but at the moment let me note only that so great is the admiration that Euclid commands that, in 1923, the poet Edna St. Vincent Millay wrote a poem entitled "Euclid Alone Has Looked on Beauty Bare." Actually, even Millay's annotated notebook from her course in Euclidean geometry has been preserved. Euclid defined a proportion derived from a simple division of a line into what he called its "extreme and mean ratio." In Euclid's words:

> A straight line is said to have been cut in extreme and mean ratio
> when, as the whole line is to the greater segment, so is the greater
> to the lesser.

A ——————————————————— C ———————— B

Figure 2

In other words, if we look at Figure 2, line *AB* is certainly longer than the segment *AC*; at the same time, the segment *AC* is longer than *CB*.

If the ratio of the length of *AC* to that of *CB* is the same as the ratio of *AB* to *AC,* then the line has been cut in extreme and mean ratio, or in a Golden Ratio.

Who could have guessed that this innocent-looking line division, which Euclid defined for some purely geometrical purposes, would have consequences in topics ranging from leaf arrangements in botany to the structure of galaxies containing billions of stars, and from mathematics to the arts? The Golden Ratio therefore provides us with a wonderful example of that feeling of utter amazement that the famous physicist Albert Einstein (1879–1955) valued so much. In Einstein's own words: "The fairest thing we can experience is the mysterious. It is the fundamental emotion which stands at the cradle of true art and science. He who knows it not and can no longer wonder, no longer feel amazement, is as good as dead, a snuffed-out candle."

As we shall see calculated in this book, the precise value of the Golden Ratio (the ratio of *AC* to *CB* in Figure 2) is the never-ending, never-repeating number 1.6180339887 . . . , and such never-ending numbers have intrigued humans since antiquity. One story has it that when the Greek mathematician Hippasus of Metapontum discovered, in the fifth century B.C., that the Golden Ratio is a number that is neither a whole number (like the familiar 1, 2, 3, . . .) nor even a ratio of two whole numbers (like the fractions ½, ⅔, ¾, . . . ; known collectively as *rational numbers*), this absolutely shocked the other followers of the famous mathematician Pythagoras (the Pythagoreans). The Pythagorean worldview (which will be described in detail in Chapter 2) was based on an extreme admiration for the *arithmos*—the intrinsic properties of whole numbers or their ratios—and their presumed role in the cosmos. The realization that there exist numbers, like the Golden Ratio, that go on forever without displaying any repetition or pattern caused a true philosophical crisis. Legend even claims that, overwhelmed with this stupendous discovery, the Pythagoreans sacrificed a hundred oxen in awe, although this appears highly unlikely, given the fact that the Pythagoreans were strict vegetarians. I should emphasize at this point that many of these stories are based on poorly documented historical material. The precise date for the discovery of numbers that are neither whole nor fractions, known as *irrational numbers,* is not known with any certainty. Nevertheless, some researchers do place the discovery in the fifth century B.C., which is at least consistent with the dating of the stories just described. What is clear is that the Pythagoreans basically believed that the existence of such numbers was so horrific that it must represent some sort of cosmic error, one that should be suppressed and kept secret.

The fact that the Golden Ratio cannot be expressed as a fraction (as a rational number) means simply that the ratio of the two lengths *AC* and *CB* in Figure 2 cannot be expressed as a fraction. In other words, no matter how hard we search, we cannot find some common measure that is contained, let's say, 31 times in *AC* and 19 times in *CB*. Two such lengths that have no common measure are called *incommensurable.* The discovery that the Golden Ratio is an irrational number was therefore, at the same time, a discovery of incommensurability. In *On the Pythagorean Life* (ca. A.D. 300), the philosopher and historian Iamblichus, a descendant of a noble Syrian family, describes the violent reaction to this discovery:

> They say that the first [human] to disclose the nature of commensurability and incommensurability to those unworthy to share in the theory was so hated that not only was he banned from [the Pythagoreans'] common association and way of life, but even his tomb was built, as if [their] former colleague was departed from life among humankind.

In the professional mathematical literature, the common symbol for the Golden Ratio is the Greek letter tau (τ; from the Greek τομή, to-mi', which means "the cut" or "the section"). However, at the beginning of the twentieth century, the American mathematician Mark Barr gave the ratio the name of phi (φ), the first Greek letter in the name of Phidias, the great Greek sculptor who lived around 490 to 430 B.C. Phidias' greatest achievements were the "Athena Parthenos" in Athens and the "Zeus" in the temple of Olympia. He is traditionally also credited with having been in charge of other Parthenon sculptures, although it is quite probable that many were actually made by his students and assistants. Barr decided to honor the sculptor because a number of art historians maintained that Phidias had made frequent and meticulous use of the Golden Ratio in his sculpture. (We shall examine similar claims very scrupulously in this book.) I will use the names Golden Ratio, Golden Section, Golden Number, phi, and also the symbol φ interchangeably throughout, because these are the names most frequently encountered in the recreational mathematics literature.

Some of the greatest mathematical minds of all ages, from Pythagoras and Euclid in ancient Greece, through the medieval Italian mathematician Leonardo of Pisa and the Renaissance astronomer Johannes Kepler, to present-day scientific figures such as Oxford physicist Roger Penrose, have spent endless hours over this simple ratio and its properties. But the fascination with the Golden Ratio is not confined just to mathematicians. Biologists, artists, musicians, historians, architects, psychologists, and even mystics have pondered and debated the basis of its ubiquity and appeal. In fact, it is probably fair to say that the Golden Ratio has inspired thinkers of all disciplines like no other number in the history of mathematics.

An immense amount of research, in particular by the Canadian mathematician and author Roger Herz-Fischler (described in his excellent book *A Mathematical History of the Golden Number*), has been devoted even just to the simple question of the origin of the name "Golden Section." Given the enthusiasm that this ratio has generated since antiquity, we might have thought that the name also has ancient origins. Indeed, some authoritative books on the history of mathematics, like François Lasserre's *The Birth of Mathematics in the Age of Plato,* and Carl B. Boyer's *A History of Mathematics,* place the origin of this name in the fifteenth and sixteenth centuries, respectively. This, however, appears not to be the case. As far as I can tell from reviewing much of the historical fact-finding effort, this term was first used by the German mathematician Martin Ohm (brother of the famous physicist Georg Simon Ohm, after whom Ohm's law in electromagnetism is named), in the 1835 second edition of his book *Die Reine Elementar-Mathematik* (The pure elementary mathematics). Ohm writes in a footnote: "One also customarily calls this division of an arbitrary line in two such parts the golden section." Ohm's language clearly leaves us with the impression that he did not invent the term himself but rather used a commonly accepted name. Yet the fact that he did not use it in the first edition of his book (published in 1826) suggests at least that the name "Golden Section" (or, in German, "Goldene Schnitt") gained its popularity only around the 1830s. The name might have been used orally prior to that, perhaps in nonmathematical circles. There is no question, however, that following Ohm's book, the term "Golden Section" started to appear frequently and repeatedly in the German mathematical and art history literature. It may have made its debut in English in an article by James Sully on aesthetics, which appeared in the ninth edition of the *Encyclopaedia Britannica* in 1875. Sully refers to the "interesting experimental enquiry . . . instituted by [Gustav Theodor] Fechner [a physicist and pioneering German psychologist in the nineteenth century] into the alleged superiority of 'the golden section' as a visible proportion." (I discuss Fechner's experiments in Chapter 7.) The earliest English uses in a mathematical context appear to have been in an article entitled "The Golden Section" (by E. Ackermann) that appeared in 1895 in the *American Mathematical Monthly* and, around the same time, in the 1898 book *Introduction to Algebra* by the well-known teacher and author G. Chrystal (1851–1911). Just as a curiosity, let me note that the only definition of a "Golden Number" that appears in the 1900 edition of the French encyclopedia *Nouveau Larousse Illustré* is: "A number used to indicate each of the years of the lunar cycle." This refers to the position of a calendar year within the nineteen-year cycle after which the phases of the Moon recur on the same dates. Clearly the phrase took a longer time to enter the French mathematical nomenclature.

But what is all the fuss about? What is it that makes this number, or geometrical proportion, so exciting as to deserve all of this attention?

The Golden Ratio's attractiveness stems first and foremost from the fact that it has an almost uncanny way of popping up where it is least expected.

Take, for example, an ordinary apple, the fruit often associated (probably mistakenly) with the tree of knowledge that figures so prominently in the biblical account of humankind's fall from grace, and cut it through its girth. You will find that the apple's seeds are arranged in a five-pointed star pattern, or pentagram (Figure 3). Each of the five

Figure 3

isosceles triangles that make the corners of a penta-gram has the property that the ratio of the length of its longer side to the shorter one (the implied base) is equal to the Golden Ratio, 1.618. . . . But, you may think, maybe this is not so surprising. After all, since the Golden Ratio has been defined as a geometrical proportion, perhaps we should not be too astonished to discover that this proportion is found in some geometrical shapes.

This is, however, only the tip of the iceberg. According to Buddhist tradition, in one of Buddha's sermons he did not utter a single word; he merely held a flower in front of his audience. What can a flower teach us? A rose, for example, is often taken as a symbol of natural symmetry, harmony, love, and fragility. In *Religion of Man,* Indian poet and philoso-pher Rabindranath Tagore (1861–1941) writes: "Somehow we feel that through a rose the language of love reached our hearts." Suppose you want to quantify the symmetric appearance of a rose. Take a rose and dis-sect it, to uncover the way in which its petals overlap their predecessors. As I describe in Chapter 5, you will find that the positions of the petals are arranged according to a mathematical rule that relies on the Golden Ratio.

Turning now to the animal kingdom, we are all familiar with the strikingly beautiful spiral structures of many shells of mollusks, such as the chambered nautilus *(Nautilus pompilius;* Figure 4). In fact, the dancing Shiva of the Hindu myth holds such a nautilus in one of his hands, as a symbol of one of the instruments initiat-ing creation. These shells also have in-spired many architectural constructions. American architect Frank Lloyd Wright (1869–1959), for example, based the de-sign of the Guggenheim Museum in New York City on the structure of the chambered nautilus. Within the mu-seum, the visitors ascend a spiral ramp, moving on, when their imaginative ca-pacity is saturated by the art they see,

Figure 4

just as the mollusk builds its spiral chambers when fully occupying its physical space. We shall discover in Chapter 5 that the growth of spiral shells also obeys a pattern that is governed by the Golden Ratio.

Figure 5

By now, we do not have to be number mysticists to begin to feel a certain awe at this property of the Golden Ratio to show up in what appear to be totally unrelated situations and phenomena. Furthermore, as I noted at the beginning of this chapter, the Golden Ratio can be found not only in natural phenomena but also in a variety of human-made objects and works of art. For example, in Salvador Dali's painting from 1955, "Sacrament of the Last Supper" (in the National Gallery, Washington D.C.; Figure 5), the dimensions of the painting (approximately 105½" × 65¾") are in a Golden Ratio to each other. Perhaps even more important, part of a huge dodecahedron (a twelve-faced regular solid in which each side is a pentagon) is seen floating above the table and engulfing it. As we shall see in Chapter 4, regular solids (like the cube) that can be precisely enclosed by a sphere (with all their corners resting on the sphere), and the dodecahedron in particular, are intimately related to the Golden Ratio. Why did Dali choose to exhibit the Golden Ratio so prominently in this painting? His remark that "the Communion must be symmetrical" only begins to answer this question. As I show in Chapter 7, the Golden Ratio features (or is at least claimed to feature) in the works of many other artists, architects, and designers, and even in famous musical compositions. Broadly speaking, the Golden Ratio has been used in some of these works to achieve what we might term "visual (or audio) effectiveness." One of the properties contributing to such effectiveness is *proportion*—the size relationships of parts to one another and to the whole. The history of art shows that in the long search for an elusive canon of "perfect" proportion, one that would somehow automatically confer aesthetically pleasing qualities on all works of art, the Golden Ratio has proven to be the most enduring. But why?

A closer examination of the examples from nature and from the arts reveals that they raise questions at three different levels of increasing depth. First, there are the immediate questions: (a) Are all the appearances of phi in nature and in the arts that are cited in the literature real, or do some of those simply represent misconceptions and crankish interpretations? (b) Can we actually explain the appearance (if real) of phi in these and other circumstances? Second, given that we define "beauty," as, for example, in *Webster's Unabridged Dictionary,* "the quality which makes an object seem pleasing or satisfying in a certain way," this raises the question: Is there an aesthetic component to mathematics? And if so, what is the essence of this component? This is a serious question because, as the American architect, mathematician, and engineer Richard Buckminster Fuller (1895–1983) once put it: "When I am working on a problem, I never think about beauty. I think only of how to solve the problem. But when I have finished, if the solution is not beautiful, I know it is wrong." Finally, the most intriguing question is: What is it that makes mathematics so powerful and ubiquitous? What is the reason that mathematics and numerical constants like the Golden Ratio play such a central role in topics ranging from fundamental theories of the universe to the stock market? Does mathematics exist even independently of the humans who have discovered/invented it and its principles? Is the universe by its very nature mathematical? This last question can be rephrased, using a famous aphorism of the British physicist Sir James Jeans (1847–1946), as: Is God a mathematician?

I will attempt to address all of these questions in some detail in this book, via the fascinating story of phi. The sometimes-tangled history of this ratio spans millennia as well as continents. Equally important, I hope to tell a good human-interest story. A part of this story will be about a time when "scientists" and "mathematicians" were self-selected individuals who simply pursued questions that kindled their curiosity. These people often labored and died without knowing whether their works would change the course of scientific thought or would simply disappear without a trace.

Before we embark on this main journey, however, we have to familiarize ourselves with numbers in general and with the Golden Ratio in particular. After all, how did the initial idea of the Golden Ratio arise? What was it that led Euclid even to bother to define such a line division? My aim is to help you glean some insights into the true roots of what we might call Golden Numberism. To this goal, we will now take a brief exploratory tour through the very dawn of mathematics.

2

THE PITCH AND THE
PENTAGRAM

*As far as the laws of mathematics refer to reality, they are not certain;
and as far as they are certain, they do not refer to reality.*
—ALBERT EINSTEIN (1879–1955)

*I see a certain order in the universe and math is
one way of making it visible.*
—MAY SARTON (1912–1995)

No one knows for sure when humans started to count, that is, to measure multitude in a quantitative way. In fact, we do not even know with certainty whether numbers like "one," "two," "three" (the cardinal numbers) preceded numbers like "first," "second," "third" (the ordinal numbers), or vice versa. Cardinal numbers simply determine the plurality of a collection of items, such as the number of children in a group. Ordinal numbers, on the other hand, specify the order and succession of specific elements in a group, such as a given date in a month or a seat number in a concert hall. Originally it was assumed that counting developed specifically to address simple day-to-day needs, which clearly argued for cardinal numbers appearing first. However, some anthropologists have suggested that numbers may have first appeared on the historical scene in relation to some rituals that required the successive appearance (in a specified order) of individuals during ceremonies. If true, this idea suggests that the ordinal number concept may have preceded the cardinal one.

Clearly, an even bigger mental leap was required to move from the simple counting of objects to an actual understanding of numbers as abstract quantities. Thus, while the first notions of numbers might have been related primarily to *contrasts,* associated perhaps with survival—Is it *one* wolf or a *pack* of wolves?—the actual understanding that two hands and two nights are both manifestations of the number 2 probably took centuries to grasp. The process had to go through the recognition of similarities (as opposed to contrasts) and correspondences. Many languages contain traces of the original divorce between the simple act of counting and the abstract concept of numbers. In the Fiji Islands, for example, the term for ten coconuts is "koro," while for ten boats it is "bolo." Similarly, among the Tauade in New Guinea, different words are used for talking about pairs of males, pairs of females, and mixed pairs. Even in English, different names often are associated with the same numbers of different aggregations. We say "a yoke of oxen" but never "a yoke of dogs."

Surely the fact that humans have as many hands as they have feet, eyes, or breasts helped in the development of the abstract understanding of the number 2. Even there, however, it must have taken longer to associate this number with things that are not identical, such as the fact that there are two major lights in the heavens, the Sun and the Moon. There is little doubt that the first distinctions were made between one and two and then between two and "many." This conclusion is based on the results of studies conducted in the nineteenth century among populations that were relatively unexposed to mainstream civilization and on linguistic differences in the terms used for different numbers in both ancient and modern times.

THREE IS A CROWD

The first indication of the fact that numbers larger than two were once treated as "many" comes from some five millennia ago. In the language of Sumer in Mesopotamia, the name for the number 3, "es," served also as the mark of plurality (like the suffix *s* in English). Similarly, ethnographic studies in 1890 of the natives of the islands in the Torres Strait, between Australia and Papua New Guinea, showed that they used a system known as two-counting. They used the words "urapun" for "one," "okosa" for "two," and then combinations such as "okosa-urapun" for "three" and "okosa-okosa" for "four." For numbers larger than four, the islanders used the word "ras" (many). Almost identical forms of nomenclatures were found in other indigenous populations from Brazil (the Botocudos) to South Africa (Zulus). The Aranda of Australia, for example, had "ninta" for "one," "tara" for "two," and then "tara mi ninta" for "three" and "tara ma tara" for "four," with all other numbers expressed as "many." Many of these populations were also found to have the tendency to group things in pairs, as opposed to counting them individually.

An interesting question is: Why did the languages used in these and other counting systems evolve to "four" and then stop (even though three and four were already expressed in terms of one and two)? One explanation suggests that this may simply reflect the fact that we happen to have four fingers in a similar position on our hands. Another, more subtle idea proposes that the answer lies in a physiological limit on human visual perception. Many studies show that the largest number we are able to capture at a glance, *without counting,* is about four or five. You may remember that in the movie *Rain Man,* Dustin Hoffman plays an autistic person with an unusual (in fact, highly exaggerated) perception of and memory for numbers. In one scene, all the toothpicks but four from a toothpick box scatter all over the floor, and he is able to tell at a glance that there are 246 toothpicks on the floor. Well, most people are unable to perform such feats. Anyone who ever tried to tally votes of any kind is familiar with this fact. We normally record the first four votes as straight lines, and then we cross those with a fifth line once a fifth vote is cast, simply because of the difficulty to perceive at a glance any number of lines that is larger than four. This system has been known in English pubs (where the barman counts the beers ordered) as the five barred gate. Curiously, an experiment described by the historian of mathematics Tobias Dantzig (1884–1956) in 1930 (in his wonderful book *Number, the Language of Science*) suggests that some birds also can recognize and discriminate among up to four objects. Dantzig's story goes as follows:

> A squire was determined to shoot a crow which made its nest in the watch-tower of his estate. Repeatedly he had tried to surprise the bird, but in vain: at the approach of man the crow would leave its nest. From a distant tree it would watchfully wait until the man had left the tower and then return to its nest. One day the squire hit upon a ruse: two men entered the tower, one remained within, the other came out and went on. But the bird was not deceived: it kept away until the man within came out. The experiment was repeated on the succeeding days with two, three, then four men, yet without success. Finally, five men were sent: as before, all entered the tower, and one remained while the other four came out and went away. Here the crow lost count. Unable to distinguish between four and five it promptly returned to its nest.

More pieces of evidence suggest that the initial counting systems followed the "one, two, . . . many" philosophy. These come from linguistic differences in the treatments of plurals and of fractions. In Hebrew, for example, there is a special form of plural for some pairs of identical items (e.g., hands, feet) or for words representing objects that contain two identical parts (e.g., pants, eyeglasses, scissors) that is different from the normal plural. Thus, while normal plurals end in "im" (for items considered masculine) or "ot" (for feminine items), the plural form for eyes, breasts, and so on, or the words for objects with two identical parts, end in "ayim." Similar forms exist in Finnish and used to exist (until medieval times) in Czech. Even more important, the transition to fractions, which surely required a higher degree of familiarity with numbers, is characterized by a marked linguistic difference in the names of fractions other than a half. In Indo-European languages, and even in some that are not (e.g., Hungarian and Hebrew), the names for the fractions "one-third" (⅓), "one-fifth" (⅕), and so on generally derive from the names of the numbers of which these fractions are reciprocals (three, five, etc.). In Hebrew, for example, the number "three" is "shalosh" and "one-third" is "shlish." In Hungarian "three" is "Hàrom" and "one-third" is "Harmad." This is not true, however, for the number "half," which is not related to "two." In Romanian, for example, "two" is "doi" and "half" is "jumate"; in Hebrew "two" is "shtayim" and "half" is "hetsi"; in Hungarian "two" is "kettö" and "half" is "fél." The implication may be that while the number ½ was understood relatively early, the notion and comprehension of other fractions as reciprocals (namely, "one over") of integer numbers probably developed only after counting passed the "three is a crowd" barrier.

COUNTING MY NUMBERLESS FINGERS

Even before the counting systems truly developed, humans had to be able to record some quantities. The oldest archaeological records that are believed to be associated with counting of some sort are in the form of bones on which regularly spaced incisions have been made. The earliest, dating to about 35,000 B.C., is a part of a baboon's thigh bone found in a cave in the Lembedo Mountains in Africa. That bone was engraved with twenty-nine incisions. Another such "bookkeeping" record, a bone of a wolf with fifty-five incisions (twenty-five in one series and thirty in another, the first series grouped in fives), was found by archaeologist Karel Absolon in 1937 at Dolné Věstonice, Czechoslovakia, and has been dated to the Aurignacian era (about 30,000 years ago). The grouping into 5, in particular, suggests the concept of a *base,* which I will discuss shortly. While we do not know the exact purpose of these incisions, they may have served as a record of a hunter's kills. The grouping would have helped the hunter to keep tally without having to recount every notch. Similarly marked bones, from the Magdalenian era (about 15,000 years ago), were also found in France and in the Pekarna cave in the Czech Republic.

A bone that has been subjected to much speculation is the Ishango bone found by archaeologist Jean de Heinzelin at Ishango near the border between Uganda and Zaire (Figure 6). That bone handle of a tool, dating to about 9000 B.C., displays three rows of notches arranged, respectively, in the following groups: (i) 9, 19, 21, 11; (ii) 19, 17, 13,

11; (iii) 7, 5, 5, 10, 8, 4, 6, 3. The sum of the numbers in the first two rows is 60 in each, which led some to speculate that they may represent a record of the phases of the Moon in two lunar months (with the possibility that some incisions may have been erased from the third row, which adds up only to 48). More intricate (and far more speculative) interpretations also have been proposed. For example, on the basis of the fact that the second row (19, 17, 13, 11) contains sequential primes (numbers that have no divisors except for 1 and the number itself), and the first row (9, 19, 21, 11) contains numbers that are different by 1 from either 10 or 20, de Heinzelin concluded that the Ishango people had some rudimentary knowledge of arithmetic and even of prime numbers. Needless to say, many researchers find this interpretation somewhat far-fetched.

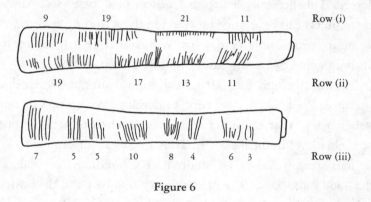

Figure 6

The Middle East has produced another interesting recording system, dating to the period between the ninth and second millennia B.C. In places ranging from Anatolia in the north to Sudan in the south, archaeologists have discovered hoards of little toylike objects of different shapes made of clay. They are in the form of disks, cones, cylinders, pyramids, animal shapes, and others. University of Texas at Austin archaeologist Denise Schmandt-Besserat, who studied these objects in the late 1970s, developed a fascinating theory. She believes that these clay objects served as pictogram tokens in the market, symbolizing the types of objects being counted. Thus, a small clay sphere might have stood for some quantity of grain, a cylinder for a head of cattle, and so on. The mideastern prehistoric merchants could therefore, according to

Schmandt-Besserat's hypothesis, conduct the accounting of their business by simply lining up the tokens according to the types of goods being transacted.

Whatever type of symbols was used for different numbers—incisions on bones, clay tokens, knots on strings (devices called quipu, used by the Inca), or simply the fingers—at some point in history humans faced the challenge of being able to represent and manipulate large numbers. For practical reasons, no symbolic system that has a uniquely different name or different representing object for every number can survive for long. In the same way that the letters in the alphabet represent in some sense the minimal number of characters with which we can express our entire vocabulary and all written knowledge, a minimal set of symbols with which all the numbers can be characterized had to be adopted. This necessity led to the concept of a "base" set—the notion that numbers can be arranged hierarchically, according to certain units. We are so familiar in everyday life with base 10 that it is almost difficult to imagine that other bases could have been chosen.

The idea behind base 10 is really quite simple, which does not mean it did not take a long time to develop. We group our numbers in such a way that ten units at a given level correspond to one unit at a higher level in the hierarchy. Thus 10 "ones" correspond to 1 "ten," 10 "tens" correspond to 1 "hundred," 10 "hundreds" correspond to 1 "thousand," and so on. The names for the numbers and the positioning of the digits also reflect this hierarchical grouping. When we write the number 555, for example, although we repeat the same cipher three times, it means something different each time. The first digit from the right represents 5 units, the second represents 5 tens, or 5 times ten, and the third 5 hundreds, or 5 times ten squared. This important rule of position, the *place-value system,* was first invented by the Babylonians (who used 60 as their base, as described below) around the second millennium B.C., and then, over a period of some 2,500 years, was reinvented, in succession, in China, by the Maya in Central America, and in India.

Of all Indo-European languages, Sanskrit, originating in northern India, provides some of the earliest written texts. In particular, four of the ancient scriptures of Hinduism, all having the Sanskrit word "veda" (knowledge) in their title, date to the fifth century B.C. The numbers 1 to 10 in Sanskrit all have different names: eka, dvau, trayas, catvaras, pañca, ṣaṭ, sapta, aṣṭau, náva, daśa. The numbers 11 to 19 are all simply a combination of the number of units and 10. Thus, 15 is "pañca-daśa," 19 is "náva-daśa," and so on. English, for example, has the equivalent "teen" numbers. In case you wonder, by the way, where "eleven" and "twelve" in English came from, "eleven" derives from "an" (one) and "lif" (left, or remainder) and "twelve" from "two" and "lif" (two left). Namely, these numbers represent "one left" and "two left" after ten. Again as in English, the Sanskrit names for the tens ("twenty," "thirty," etc.) express the unit and plural tens (e.g., 60 is ṣaṣti), and all Indo-European languages have a very similar structure in their vocabulary for numbers. So the users of these languages quite clearly adopted the base 10 system.

There is very little doubt that the almost universal popularity of base 10 stems simply from the fact that we happen to have ten fingers. This possibility was already raised by the Greek philosopher Aristotle (384–322 B.C.) when he wondered (in *Problemata*): "Why do all men, barbarians and Greek alike, count up to ten and not up to any other number?" Base 10 really offers no other superiority over, say, base 13. We could even argue theoretically that the fact that 13 is a prime number, divisible only by 1 and itself, gives it an advantage over 10, because most fractions would be irreducible in such a system. While, for example, under base 10 the number $^{36}/_{100}$ also can be expressed as $^{18}/_{50}$ or $^{9}/_{25}$, such multiple representations would not exist under a prime base like 13. Nevertheless, base 10 won, because ten fingers stood out in front of every human's eyes, and they were easy to use. In some Malay-Polynesian languages, the word for "hand," "lima," is actually the same as the word for "five." Does this mean that all the known civilizations chose 10 as their base? Actually, no.

Of the other bases that have been used by some populations around the world, the most common was base 20, known as the vigesimal base. In this counting system, which was once popular in large portions of Western Europe, the grouping is based on 20 rather than 10. The choice of this system almost certainly comes from combining the fingers with the toes to form a larger base. For the Inuit (Eskimo) people, for example, the number "twenty" is expressed by a phrase with the meaning "a man is complete." A number of modern languages still have traces of a vigesimal base. In French, for example, the number 80 is "quatre-vingts" (meaning "four twenties"), and an archaic form of "six-vingts" ("six twenties") existed as well. An even more extreme example is provided by a thirteenth-century hospital in Paris, which is still called L'Ôpital de Quinze-Vingts (The Hospital of Fifteen Twenties), because it was originally designed to contain 300 beds for blind veterans. Similarly, in Irish, 40 is called "daichead," which is derived from "da fiche" (meaning "two times twenty"); in Danish, the numbers 60 and 80 ("tresindstyve" and "firsindstyve" respectively, shortened to "tres" and "firs") are literally "three twenties" and "four twenties."

Probably the most perplexing base found in antiquity, or at any other time for that matter, is base 60—the sexagesimal system. This was the system used by the Sumerians in Mesopotamia, and even though its origins date back to the fourth millennium B.C., this division survived to the present day in the way we represent time in hours, minutes, and seconds as well as in the degrees of the circle (and the subdivision of degrees into minutes and seconds). Sixty as a base for a number system taxes the memory considerably, since such a system requires, in principle, a unique name or symbol for all the numbers from 1 to 60. Aware of this difficulty, the ancient Sumerians used a certain trick to make the numbers easier to remember—they inserted 10 as an intermediate step. The introduction of 10 allowed them to have unique names for the numbers 1 to 10; the numbers 10 to 60 (in units of 10) represented combinations of names. For example, the Sumerian word for 40, "nišmin," is a combination of the word for 20, "niš," and the word for 2, "min." If we write the number 555 in a purely sexagesimal system, what we mean is $5 \times (60)^2 + 5 \times (60) + 5$, or 18,305 in our base 10 notation.

Many speculations have been advanced as to the logic or circumstances that led the Sumerians to choose the unusual base of 60. Some are based on the special mathematical properties of the number 60: It is the first number that is divisible by 1, 2, 3, 4, 5, and 6. Other hypotheses attempt to relate 60 to concepts such as the number of months in a year or days in a year (rounded to 360), combined somehow with the numbers 5 or 6. Most recently, French math teacher and author Georges Ifrah argued in his superb 2000 book, *The Universal History of Numbers,* that the number 60 may have been the consequence of the mingling of two immigrant populations, one of which used base 5 and the other base 12. Base 5 clearly originated from the number of fingers on one hand, and traces for such a system can still be found in a few languages, such as in the Khmer in Cambodia and more prominently in the Saraveca in South America. Base 12, for which we find many vestiges even today—for example, in the British system of weights and measures—may have had its origins in the number of joints in the four fingers (excluding the thumb; the latter being used for the counting).

Incidentally, strange bases pop up in the most curious places. In Lewis Carroll's *Alice's Adventures in Wonderland,* to assure herself that she understands the strange occurrences around her, Alice says: "I'll try if I know all the things I used to know. Let me see: four times five is twelve, and four times six is thirteen, and four times seven is—oh dear! I shall never get to twenty at that rate!" In his notes to Carroll's book, *The Annotated Alice,* the famous mathematical recreation writer Martin Gardner provides a nice explanation for Alice's bizarre multiplication table. He proposes that Alice is simply using bases other than 10. For example, if we use base 18, then $4 \times 5 = 20$ will indeed be written as 12, because 20 is 1 unit of 18 and 2 units of 1. What lends plausibility to this explanation is of course the fact that Charles Dodgson ("Lewis Carroll" was his pen name) lectured on mathematics at Oxford.

OUR NUMBERS, OUR GODS

Irrespective of the base that any of the ancient civilizations chose, the first group of numbers to be appreciated and understood at some level was the group of whole numbers (or *natural* numbers). These are the familiar 1, 2, 3, 4, . . . Once humans absorbed the comprehension of these numbers as abstract quantities into their consciousness, it did not take them long to start to attribute special properties to numbers. From Greece to India, numbers were accredited with secret qualities and powers. Some ancient Indian texts claim that numbers are almost divine, or "Brahma-natured." These manuscripts contain phrases that are nothing short of worship to numbers (like "hail to one"). Similarly, a famous dictum of the Greek mathematician Pythagoras (whose life and work will be described later in this chapter) suggests that "everything is arranged according to number." These sentiments led on one hand to important developments in number theory but, on the other, to the development of *numerology*—the set of doctrines according to which all aspects of the universe are associated with numbers and their idiosyncrasies. To the numerologist, numbers were fundamental realities, drawing symbolic meanings from the relation between the heavens and human activities. Furthermore, essentially no number that was mentioned in the holy writings was ever treated as irrelevant. Some forms of numerology affected entire nations. For example, in the year 1240 Christians and Jews in Western Europe expected the arrival of some messianic king from the East, because it so happened that the year 1240 in the Christian calendar corresponded to the year 5000 in the Jewish calendar. Before we dismiss these sentiments as romantic naïveté that could have happened only many centuries ago, we should recall the extravagant hoopla surrounding the ending of the last millennium.

One special version of numerology is the Jewish Gematria (possibly based on "geometrical number" in Greek), or its Muslim and Greek analogues, known as Khisab al Jumal ("calculating the total"), and Isopsephy (from the Greek "isos," equal, and "psēphizein," to count), respectively. In these systems, numbers are assigned to each letter of the alphabet of a language (usually Hebrew, Greek, Arabic, or Latin). By adding together the values of the constituent letters, numbers are then associated with words or even entire phrases. Gematria was especially popular in the system of Jewish mysticism practiced mainly from the thirteenth to the eighteenth century known as cabala. Hebrew scholars sometimes used to amaze listeners by calling out a series of apparently random numbers for some ten minutes and then repeating the series without an error. This feat was accomplished simply by translating some passage of the Hebrew scriptures into the language of Gematria.

One of the most famous examples of numerology is associated with 666, the "number of the Beast." The "Beast" has been identified as the Antichrist. The text in the Book of Revelations (13:18) reads: "This calls for wisdom: let anyone with understanding calculate the number of the beast, for it is the number of a man. Its number is six hundred and sixty-six." The phrase "it is the number of a man" prompted many of the Christian mystics to attempt to identify historical figures whose names in Gematria or Isopsephy give the value 666. These searches led to, among others, names like those of Nero Caesar and the emperor Diocletian, both of whom persecuted Christians. In Hebrew, Nero Caesar was written as (from right to left): נרון קסר, and the numerical values assigned in Gematria to the Hebrew letters (from right to left)—50, 200, 6, 50; 100, 60, 200—add up to 666. Similarly, when only the letters that are also Roman numerals (D, I, C, L, V) are counted in the Latin name of Emperor Diocletian, DIOCLES AVGVSTVS, they also add up to 666 (500 + 1 + 100 + 50 + 5 + 5 + 5). Clearly, all of these associations are not only fanciful but also rather contrived (e.g., the spelling in Hebrew of the word Caesar actually omits a letter, of value 10, from the more common spelling).

Amusingly, in 1994, a relation was "discovered" (and appeared in the *Journal of Recreational Mathematics*) even between the "number of the Beast" and the Golden Ratio. With a scientific pocket calculator, you can use the trigonometric functions sine and cosine to calculate the value of the expression [sin 666° + cos (6 × 6 × 6)°]. Simply enter 666 and hit the [sin] button and save that number, then enter 216 (= 6 × 6 × 6) and hit the [cos] button, and add the number you get to the number you saved. The number you will obtain is a good approximation of the negative of phi. Incidentally, President Ronald Reagan and Nancy Reagan changed their address in California from 666 St. Cloud Road to 668 to avoid the number 666, and 666 was also the combination to the mysterious briefcase in Quentin Tarantino's movie *Pulp Fiction*.

One clear source of the mystical attitude toward whole numbers was the manifestation of such numbers in human and animal bodies and in the cosmos, as perceived by the early cultures. Not only do humans have the number 2 exhibited all over their bodies (eyes, hands, nostrils, feet, ears, etc.), but there are also two genders, there is the Sun-Moon system, and so on. Furthermore, our subjective time is divided into three tenses (past, present, future), and, due to the fact that Earth's rotation axis remains more or less pointed in the same direction (roughly toward the North Star, Polaris, although small variations do exist, as described in Chapter 3), the year is divided into four seasons. The seasons simply reflect the fact that the orientation of Earth's axis relative to the Sun changes over the course of the year. The more directly a part of the Earth is exposed to sunlight, the longer the daylight hours and the warmer the temperature. In general, numbers acted in many circumstances as the mediators between cosmic phenomena and human everyday life. For example, the names of the seven days of the week were based on the names of the celestial objects originally considered to be planets: the Sun, the Moon, Mars, Mercury, Jupiter, Venus, and Saturn.

The whole numbers themselves are divided into odd and even, and nobody did more to emphasize the differences between the odd and even numbers, and to ascribe a whole menagerie of properties to these differences, than the Pythagoreans. In particular, we shall see that we can identify the Pythagorean fascination with the number 5 and their admiration for the five-pointed star as providing the initial impetus for the interest in the Golden Ratio.

PYTHAGORAS AND THE PYTHAGOREANS

Pythagoras was born around 570 B.C. in the island of Samos in the Aegean Sea (off Asia Minor), and he emigrated sometime between 530 and 510 to Croton in the Dorian colony in southern Italy (then known as Magna Graecia). Pythagoras apparently left Samos to escape the stifling tyranny of Polycrates (died ca. 522 B.C.), who established Samian naval supremacy in the Aegean Sea. Perhaps following the advice of his presumed teacher, the mathematician Thales of Miletus, Pythagoras probably lived for some time (as long as twenty-two years, according to some accounts) in Egypt, where he would have learned mathematics, philosophy, and religious themes from the Egyptian priests. After Egypt was overwhelmed by Persian armies, Pythagoras may have been taken to Babylon, together with members of the Egyptian priesthood. There he would have encountered the Mesopotamian mathematical lore. Nevertheless, the Egyptian and Babylonian mathematics would prove insufficient for Pythagoras' inquisitive mind. To both of these peoples, mathematics provided practical tools in the form of "recipes" designed for specific calculations. Pythagoras, on the other hand, was one of the first to grasp numbers as abstract entities that exist in their own right.

In Italy, Pythagoras began to lecture on philosophy and mathematics, quickly establishing an enthusiastic crowd of followers, which may have included the young and beautiful Theano (daughter of his host Milo), whom he later married. The atmosphere in Croton proved extremely fertile for Pythagoras' teachings, since the community there was composed of a plethora of semimystic cults. Pythagoras established a strict routine for his students, paying particular attention to the hour of waking and the hour of falling asleep. Students were advised upon rising to repeat the verses:

> *As soon as you awake, in order lay*
> *the actions to be done the coming day.*

Similarly, at nightfall, they were to recite:

> *Allow not sleep to close your eyes*
> *Before three times reflecting on*
> *Your actions of the day. What deeds*
> *Done well, what not, what left undone?*

Most of the details of Pythagoras' life and the reality of his mathematical contributions remain veiled in uncertainty. One legend has it that he had a golden birthmark on his thigh, which was taken by his followers to indicate that he was a son of the god Apollo. None of the biographies of Pythagoras written in antiquity have survived, and biographies written later, such as the *Lives of the Eminent Philosophers,* written by Diogenes Läertius in the third century, often rely on many sources of varying reliability. Pythagoras apparently wrote nothing, and yet his influence was so great that the more attentive of his followers formed a secretive society, or brotherhood, and were known as the Pythagoreans.

Aristippus of Cyrene tells us in his *Account of Natural Philosophers* that Pythagoras derived his name from the fact that he was speaking *(agoreuein)* truth like the God at Delphi *(tou Pythiou)*.

The events surrounding Pythagoras' death are as uncertain as the facts about his life. According to one story, the house in which he was staying at Croton was set on fire by a mob, envious of the Pythagorean elite, and Pythagoras himself was murdered during his escape, upon reaching a place full of beans on which he wouldn't trample. A different version is provided by the Greek scientist and philosopher Dicaearchus of Messana (ca. 355–280 B.C.), who states that Pythagoras managed to escape as far as the Temple of the Muses at Metapontum, where he died following forty days of self-imposed starvation. A completely different story is told by Hermippus, according to which Pythagoras was slain by the Syracusans in their war against the Agrigentine army, which Pythagoras joined.

Even though it is almost impossible to attribute with certainty any specific mathematical achievements either to Pythagoras himself or to his followers, there is no question that they have been responsible for a mingling of mathematics, philosophy of life, and religion unparalleled in history. In this respect it is perhaps interesting to note the historical coincidence that Pythagoras was a contemporary of Buddha and Confucius.

Pythagoras is in fact credited with having coined the words "philosophy" ("love of wisdom") and "mathematics" ("that which is learned"). To him, a "philosopher" was someone who "gives himself up to discovering the meaning and purpose of life itself . . . to uncover the secrets of nature." Pythagoras emphasized the importance of learning above all other activities, because, in his words, "most men and women, by birth or nature, lack the means to advance in wealth and power, but all have the ability to advance in knowledge." He was also famous for introducing the doctrine of metempsychosis—that the soul is immortal and is reborn or transmigrated in human and animal incarnations. This doctrine resulted in a strong advocacy of vegetarianism, since animals to be slaughtered could represent reincarnated friends. To purify the soul, the Pythagoreans established strict rules, which included, for example, a prohibition on eating beans and an extreme emphasis on the training of the memory. In his treatise *On the Pythagoreans,* the famous Greek philosopher Aristotle gives several possible reasons for the abstention from beans: They resemble genitals; being plants without parts they are like the gates of hell; beans were supposed to arise simultaneously with humans in the act of universal creation; or beans were used in elections in oligarchical governments.

Pythagoras and the Pythagoreans are best known for their presumed role in the development of mathematics and for the application of mathematics to the concept of order, whether it is musical order, the order of the cosmos, or even ethical order. Every child in school learns the Pythagorean theorem of a triangle that has a right (90-degree) angle (a right triangle). According to this theorem (Figure 7, on the right), the area of the square constructed on the longest side (the hypotenuse) equals the sum of the areas of the squares constructed on the two shorter sides. In other words, if the length of the hypotenuse is c, then the area of the square constructed on it is c^2; the areas of the squares constructed on the other two sides (of lengths a and b) are a^2 and b^2 respectively. The Pythagorean theorem can therefore be stated as: $c^2 = a^2 + b^2$ in every right triangle. In 1971, when the republic of Nicaragua selected the ten mathematical equations that changed the face of the Earth as the theme for a series of stamps, the Pythagorean theorem appeared on the second stamp. The numbers 3, 4, 5, or 7, 24, 25, for example, form Pythagorean triples, because $3^2 + 4^2 = 5^2$ (9 + 16 = 25); $7^2 + 24^2 = 25^2$ (49 + 576 = 625), and they can be used as the lengths of the sides of a right triangle.

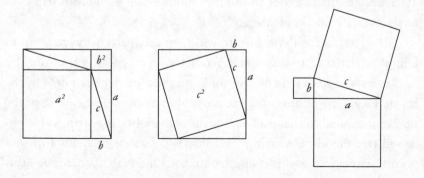

Figure 7

Figure 7 also suggests what is perhaps the easiest proof of the Pythagorean theorem: On one hand, when one subtracts from the square whose side equals $a + b$ the area of four identical triangles, one gets the square built on the hypotenuse (middle figure). On the other, when one subtracts from the same square the same four triangles in a different arrangement (left figure), one gets the two squares built on the shorter sides. Thus, the square on the hypotenuse is clearly equal in area to the sum of the two smaller squares. In his 1940 book *The Pythagorean Proposition,* mathematician Elisha Scott Loomis presented 367 proofs of the Pythagorean theorem, including proofs by Leonardo da Vinci and by the twentieth president of the United States, James Garfield.

Even though the Pythagorean theorem was not yet known as a "truth" characterizing all right-angle triangles, Pythagorean triples actually had been recognized long before Pythagoras. A Babylonian clay tablet from the Old Babylonian period (ca. 1600 B.C.) contains fifteen such triples.

The Babylonians discovered that Pythagorean triples can be constructed using the following simple procedure, or "algorithm." Choose any two whole numbers p and q such that p is larger than q. You can now form the Pythagorean triple of numbers $p^2 - q^2$; $2pq$; $p^2 + q^2$. For example, suppose q is 1 and p is 4. Then $p^2 - q^2 = 4^2 - 1^2 = 16 - 1 = 15$; $2pq = 2 \times 4 \times 1 = 8$; $p^2 + q^2 = 4^2 + 1^2 = 16 + 1 = 17$. The set of numbers 15, 8, 17 is a Pythagorean triple because $15^2 + 8^2 = 17^2$ ($225 + 64 = 289$). You can easily show that this will work for any whole numbers p and q. (For the interested reader, a brief proof is presented in Appendix 1.) Therefore, there exists an infinite number of Pythagorean triples (a fact proven by Euclid of Alexandria).

However, in the Pythagorean world, orderly patterns were far from being restricted to triangles and geometry. Pythagoras is traditionally said to have discovered the harmonic progressions in the notes of the musical scale, by finding that the musical intervals and the pitch of the notes correspond to the relative lengths of the vibrating strings. He observed that dividing a string by consecutive integers yields (up to a point) harmonious and pleasing (consonant) intervals. When two arbitrary musical notes are made to sound together, the resulting sound is usually harsh (dissonant) to our ear. Only a few combinations produce pleasant sounds. Pythagoras discovered that these rare consonances are obtained when the notes are produced by similar strings whose lengths are in ratios given by the first few whole numbers. Unison is obtained when the strings are of equal length (a 1:1 ratio); the octave is obtained by a 1:2 ratio of string lengths; the fifth by 2:3; and the fourth by 3:4. In other words, you can pluck a string and sound a note. If you pluck an equally taut string that is one-half the length, you will hear a note that is precisely one harmonic octave above the first. Similarly, ⅔ of a C-string gives the note A, ⅓ of it gives G, ½ of it gives F, and so on. These remarkable early findings formed the basis for the more advanced understanding of musical intervals that developed in the sixteenth century (in which, incidentally, Vincenzo Galilei, Galileo's father, was involved). A wonderful illustration by Franchinus Gafurius, which appeared as a frontispiece in *Theorica Musice* in 1492, shows Pythagoras experimenting with the sounds of various devices, including hammers, strings, bells, and flutes (Figure 8; the upper left depicts the biblical figure of Jubal or Tubal, "the father of all such as handle the harp and organ"). But, wondered the Pythagoreans, if musical harmony can be expressed by numbers, why not the entire cosmos? They therefore concluded that all objects in the universe owed their characteristics to the nature of number. Astronomical observations suggested, for example, that the motions in the heavens also were extremely regular and subject to a specific order. This led to the concept of a beautiful "harmony of the spheres"—the notion that in their regular motions, heavenly bodies also create harmonious music. The philosopher Porphyry (ca. A.D. 232–304), who wrote more than seventy works dealing with history, metaphysics, and literature, also wrote (as a part of his four-volume work *History of Philosophy*) a brief biography of Pythagoras entitled *Life of Pythagoras*. In it, Porphyry says about Pythagoras: "He himself could hear the harmony of the Universe, and understood the music of the spheres, and the stars which move in concert with them, and which we cannot hear because of the limitations of our weak nature." After enumerating more of Pythagoras' exquisite qualities, Porphyry continues: "Pythagoras affirmed that the Nine Muses were constituted by the sounds made by the seven planets, the sphere of the fixed stars, and that which is opposed to our earth, called the 'counter-earth' "

Figure 8

(the latter, according to the Pythagorean theory of the universe, revolved in opposition to Earth, around a central fire). The concept of the "harmony of the spheres" was elaborated upon again, more than twenty centuries later, by the famous astronomer Johannes Kepler (1571–1630). Having witnessed in his own life much agony and the horrors of war, Kepler concluded that Earth really created two notes, *mi* for misery ("miseria" in Latin) and *fa* for famine ("fames" in Latin). In Kepler's words: "the Earth sings MI FA MI, so that even from the syllable you may guess that in this home of ours Misery and Famine hold sway."

The Pythagorean obsession with mathematics was mildly ridiculed by the great Greek philosopher Aristotle. He writes in *Metaphysics* (in the fourth century B.C.): "The so-called Pythagoreans applied themselves to mathematics, and were the first to develop this science; and through penetrating it, they came to fancy that its principles are the principles of all things." Today, while we may be amused by some of the Pythagorean fanciful ideas, we have to recognize that the fundamental thought behind them is really not very different from that expressed by Albert Einstein (in *Letters to Solovine*): "Mathematics is only a means for expressing the laws that govern phenomena." Indeed, the laws of physics, sometimes referred to as the "laws of nature," simply represent mathematical formulations of the behavior that we observe all natural phenomena to obey. For example, the central idea in Einstein's theory of general relativity is that gravity is not some mysterious, attractive force that acts across space but rather a manifestation of the geometry of the inextricably linked space and time. Let me explain, using a simple example, how a geometrical property of space could be perceived as an attractive force, such as gravity. Imagine two people who start to travel precisely northward from two different points on Earth's equator. This means that at their starting points, these people travel along parallel lines (two longitudes), which, according to the plane geometry we learn in school, should never meet. Clearly, however, these two people will meet at the North Pole. If these people did not know that they were really traveling on the curved surface of a sphere, they would conclude that they must have experienced some attractive force, since they arrived at the same point in spite of starting their motions along parallel lines. Therefore, the geometrical curvature of space can manifest itself as an attractive force. The Pythagoreans were probably the first to recognize the abstract concept that the basic *forces* in the universe may be expressed through the language of mathematics.

Due perhaps to the simple harmonic ratios found in music, 1:2, 2:3, 3:4, the Pythagoreans became particularly intrigued by the differences between the odd and even numbers. They associated with the odd numbers male attributes and, rather prejudiciously, also light and goodness, while they gave the even numbers female attributes and associated with them darkness and evil. Some of these prejudices toward even and odd numbers persisted for centuries. For example, the Roman scholar Pliny the Elder, who lived A.D. 23 to 79, wrote in his *Historia Naturalis* (a thirty-seven-volume encyclopedia on natural history): "Why is it that we entertain the belief that for every purpose odd numbers are the most effectual?" Similarly, in Shakespeare's *Merry Wives of Windsor* (Act V, Scene I), Sir John Falstaff says: "They say there is divinity in odd numbers, either in nativity, chance or death." Mideastern religions produced a similar attitude. According to the Muslim tradition, the prophet Muhammad ate an odd number of dates to break his fast, and Jewish prayers often have an odd number (3, 7) of repetitions associated with them.

Besides the roles that the Pythagoreans assigned to the odd and even numbers in general, they also attributed special properties to some individual numbers. The number 1, for example, was considered the generator of all other numbers and thus not regarded as a number itself. It was also assumed to characterize reason. Geometrically, the number 1 was represented by the point, which in itself was assumed to be the generator of all dimensions. The number 2 was the first female number and also the number of opinion and of division. Somewhat similar sentiments are expressed in the yin and yang of the Chinese religious cosmology, with the yin representing the feminine and negative principle, like passivity and darkness, and the yang the bright and masculine principle. The number 2 is associated to this very day in many languages with hypocrisy and unreliability, as manifested by such expressions as "two-faced" (in Iranian) or "double-tongued" (in German and Arabic). The original identification of the number 2 with feminine and of 3 with masculine may have been inspired by the configurations of female breasts and male genitalia. This tentative conclusion is supported by the fact that the Konso of East Africa make the same identification. In everyday life, division into two categories is the most common: good and bad, up and down, right and left. Geometrically, 2 was expressed by the line (which is determined by two points), which has one dimension. Three was supposed to be the first real male number and also the number of harmony, since it combines unity (the number 1) and division (the number 2). To the Pythagoreans, 3 was in some sense the first real number because it has a "beginning," a "middle," and an "end" (unlike 2, which does not have a middle). The geometrical expression of 3 was the triangle, since three points not on the same line determine a triangle, and the area of the triangle has two dimensions.

Interestingly, 3 was also the basis for the construction of military units in the Bible. For example, in 2 Samuel 23, there is a story on the very basic unit, the "three warriors" that King David had. In the same chapter, there is a detailed count of the "thirty chiefs" who "went down to join David at the cave of Adulam," but at the end of the count the biblical editor concludes that they were "thirty-seven in all." Clearly, "thirty" was the definition of the unit, even if the actual number of members was somewhat different. In Judges 7, when Gideon needs to fight the Midianites, he chooses three hundred men, "all those who lap the water with their tongues." Moving to yet larger units, in 1 Samuel 13, "Saul chose three thousand out of Israel" to fight the Philistines, who at the same time "mustered to fight with Israel, thirty thousand chariots." Finally, in 2 Samuel 6, "David again gathered all the chosen men of Israel, thirty thousand" to fight the Philistines.

The number 4, for the Pythagoreans, was the number of justice and order. On the surface of Earth, the four winds or directions provided the necessary orientation for humans to identify their coordinates in space. Geometrically, four points that are not in the same plane can form a tetrahedron (a pyramid with four triangular faces), which has a volume in three dimensions. Another consideration that gave the number 4 a somewhat special status for the Pythagoreans was their attitude toward the number 10, or the holy *tetractys*. Ten was the most revered number, because it represented the cosmos as a whole. The fact that $1 + 2 + 3 + 4 = 10$ generated a close association between 10 and 4. At the same time, this relation meant that 10 not only united the numbers representing all dimensions but also combined all the properties of uniqueness (as expressed by 1), polarity (expressed by 2), harmony (expressed by 3), and space and matter (expressed by 4). Ten was therefore the number of *everything*, with properties best expressed by the Pythagorean Philolaus around 400 B.C.: "sublime, potent and all-creating, the beginning and the guide of the divine concerning life on Earth."

The number 6 was the first *perfect* number, and the number of creation. The adjective "perfect" was attached to numbers that are precisely equal to the sum of all the smaller numbers that divide into them, as $6 = 1 + 2 + 3$. The next such number, incidentally, is $28 = 1 + 2 + 4 + 7 + 14$, followed by $496 = 1 + 2 + 4 + 8 + 16 + 31 + 62 + 124 + 248$; by the time we reach the ninth perfect number, it contains thirty-seven digits. Six is also the product of the first female number, 2, and the first masculine number, 3. The Hellenistic Jewish philosopher Philo Judaeus of Alexandria (ca. 20 B.C.–ca. A.D. 40), whose work brought together Greek philosophy and Hebrew scriptures, suggested that God created the world in six days because six was a perfect number. The same idea was elaborated upon by St. Augustine (354–430) in *The City of God:* "Six is a number perfect in itself, and not because God created the world in six days; rather the contrary is true: God created the world in six days because this number is perfect, and it would remain perfect, even if the work of the six days did not exist." Some commentators of the Bible regarded 28 also as a basic number of the Supreme Architect, pointing to the 28 days of the lunar cycle. The fascination with perfect numbers penetrated even into Judaism, and their study was advocated in the twelfth century by Rabbi Yosef ben Yehudah Ankin in his book, *Healing of the Souls*.

I have deliberately left the number 5 for last in giving these examples of the Pythagoreans' attitude to numbers, because this number also leads us to the origins of the Golden Ratio. Five represented the union of the first female number, 2, with the first male number, 3, and as such it was the number of love and marriage. The Pythagoreans apparently used the pentagram—the five-pointed star (Figure 3)—as the symbol of their brotherhood, and they called it "Health." The second-century Greek writer and rhetorician Lucian writes (in *In Defense of a Slip of the Tongue in Greeting*):

> At any rate all his [Pythagoras'] school in serious letters to each other began straightway with "Health to you," as a greeting most suitable for both body and soul, encompassing all human goods. Indeed the Pentagram, the triple intersecting triangle which they used as a symbol of their sect, they called "Health."

An imaginative (though perhaps not altogether sound) explanation for the association of the pentagram with health was suggested by A. de la Fuÿe in his 1934 book, *Le Pentagramme Pythagoricien, Sa Diffusion, Son Emploi dans le Syllabaire Cuneiform* (The Pythagorean pentagram, its distribution, its usage in the cuneiform spelling book). De la Fuÿe proposed that the pentagram symbolized the Greek goddess of health, Hygeia, through a correspondence of the five points of the star to a cartoon-like representation of the goddess (Figure 9).

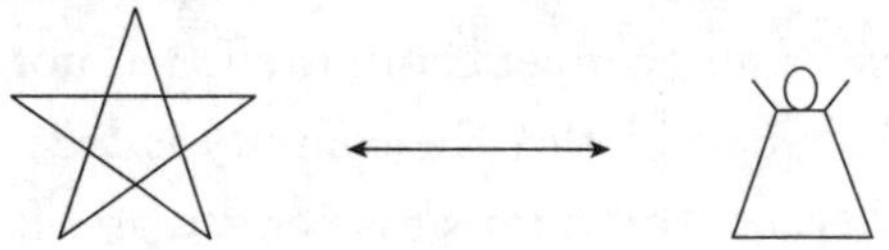

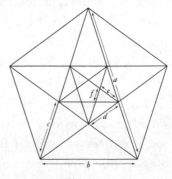

Figure 9

Figure 10

The pentagram is also closely related to the regular pentagon—the plane figure having five equal sides and equal angles (Figure 10). If you connect all the vertices of the pentagon by diagonals, you obtain a pentagram. The diagonals also form a smaller pentagon at the center, and the diagonals of this pentagon form a pentagram and a yet smaller pentagon (Figure 10). This progression can be continued ad infinitum, creating smaller and smaller pentagons and pentagrams. The striking property of all of these figures is that if you look at line segments in order of decreasing lengths (the ones marked *a, b, c, d, e, f* in the figure), you can easily prove using elementary geometry that *every segment is smaller than its predecessor by a factor that is precisely equal to the Golden Ratio, φ.* That is, the ratio of the lengths of *a* to *b* is phi; the ratio of *b* to *c* is phi; and so on. Most important, you can use the fact that the process of creating a series of nested pentagons and pentagrams can be continued indefinitely to smaller and smaller sizes to prove rigorously that the diagonal and the side of the pentagon are incommensurable, that is, the ratio of their lengths (which is equal to phi) cannot be expressed as a ratio of two whole numbers. What this means is that the diagonal and the side of the pentagon cannot have any common measure, such that the diagonal is some integer multiple of that measure and the side is also an integer multiple of the same measure. (For the more mathematically inclined reader, the proof is presented in Appendix 2.) Recall that numbers that cannot be expressed as ratios of two whole numbers (namely as fractions, or rational numbers) are known as irrational numbers. This proof therefore establishes the fact that phi is an irrational number.

Several researchers (including Kurt von Fritz in his article entitled "The Discovery of Incommensurability by Hippasus of Metapontum" published in 1945) suggested that the Pythagoreans are the ones who first discovered the Golden Ratio and incommensurability. These historians of mathematics argued that the Pythagorean preoccupation with the pentagram and the pentagon, coupled with the actual geometrical knowledge in the middle of the fifth century B.C., make it very plausible that the Pythagoreans, and in particular perhaps Hippasus of Metapontum, discovered the Golden Ratio and, through it, incommensurability. The arguments appear to be at least partially supported by the writings of the founder of the Syrian school of Neoplatonism,

Iamblichus (ca. A.D. 245–325). According to one of Iamblichus' accounts, the Pythagoreans erected a tombstone to Hippasus, as if he were dead, because of the devastating discovery of incommensurability. In another place, however, Iamblichus reports that:

> It is related of Hippasus that he was a Pythagorean, and that, owing to his being the first to publish and describe the sphere from the twelve pentagons, he perished at sea for his impiety, but he received credit for the discovery, though really it all belonged to HIM (for in this way they refer to Pythagoras, and they do not call him by his name).

In the phrase "describe the sphere from the twelve pentagons," Iamblichus refers (somewhat vaguely, since the figure is not really a sphere) to the construction of the dodecahedron, a solid with twelve faces, each of which is a pentagon, which is one of the five solids known as Platonic solids. The Platonic solids are intimately related to the Golden Ratio, and we shall return to them in Chapter 4. In spite of the somewhat mythical flavor of these accounts, the mathematical historian Walter Burkert concludes in his 1972 book *Lore and Science in Ancient Pythagoreanism* that "the tradition about Hippasus, though surrounded by legend, makes sense." The main reason for this statement is provided by Figure 10 (and Appendix 2). The conclusion that the diagonal and the side in a regular pentagon are incommensurable is based on the very simple observation that the construction of smaller and smaller pentagons can be continued indefinitely. The proof therefore would have definitely been accessible to the fifth-century B.C. mathematicians.

TO THE RATIONAL BEING ONLY THE IRRATIONAL IS UNENDURABLE

While it is certainly possible (and perhaps even likely) that incommensurability and irrational numbers were first discovered via the Golden Ratio, the more traditional view is that these concepts were discovered through the ratio of the diagonal and the side of the square. Aristotle writes in *Prior Analytics:* "the diagonal [of a square] is incommensurable [with the side] because odd numbers are equal even if it is assumed to be commensurate." Aristotle alludes here to a proof of incommensurability, which I now present in full detail, because it is a beautiful example of a proof by the logical method known as reductio ad absurdum (reduction to absurdity). In fact, when in 1988 the journal *The Mathematical Intelligencer* invited its readers to rank a selection of twenty-four theorems according to their "beauty," the proof I am about to present was ranked seventh.

The idea behind the ingenious method of reductio ad absurdum is that you prove a proposition simply by proving the falsity of its contradictory. The most influential Jewish scholar of the Middle Ages, Maimonides (Moses Ben Maimon; 1135–1204), even attempted to use this logical device to prove the existence of a creator. In his monumental work, *Mishne Torah* (The Torah reviewed), which attempts to encompass all religious subject matter, Maimonides writes: "The basic principle is that there is a First Being who brought every existing thing into being, for if it be supposed that he did not exist, then nothing else could possibly exist." In mathematics, reductio ad absurdum is used as follows. You start by assuming that the theorem you seek to prove true is in fact false. From that, by a series of logical steps you derive something that represents a clear logical *contradiction,* such as 1 = 0. You thus conclude that the original theorem could not have been false; therefore, it must be true. Note that for this method to work, you have to assume that a theorem or statement has to be *either true or false*—you are either reading this page right now or you are not.

Examine first the square in Figure 11, in which the length of the side is one unit. If we want to find the length of the diagonal, we can use the Pythagorean theorem in any of the two right triangles into which the square is divided. Recall that the theorem states that the square of the hypotenuse (the diagonal) is equal to the sum of the squares of the two shorter sides of the triangle. If we call the length of the hy-

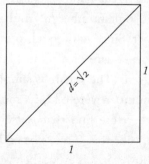

Figure 11

potenuse d, we have that: $d^2 = 1^2 + 1^2$, or $d^2 = 1 + 1 = 2$. If we know the square of a number, we find the number itself by taking the square root. For example, if we know that the square of x is equal to 25, then $x = 5 = \sqrt{25}$. From $d^2 = 2$ we therefore find that $d = \sqrt{2}$. The ratio of the diagonal to the side of a square is therefore the square root of 2. (A pocket calculator will show you that the value of the latter is equal to 1.41421356 . . . .) What we want to show now is that $\sqrt{2}$ cannot be expressed as a ratio of *any* two whole numbers (and therefore that it is an irrational number). Think about this for a moment: What we are about to prove is that even though we have an infinite collection of whole numbers at our disposal, no matter for how long we will search, we will never find two of them that have a ratio that is precisely equal to $\sqrt{2}$. Isn't this mind-boggling?

The proof (by reductio ad absurdum) goes as follows: We start by assuming the opposite of what we want to prove, namely, we assume that $\sqrt{2}$ is actually equal to some ratio of two whole numbers a and b, $\sqrt{2} = a/b$. If a and b happen to have some common factors (as the numbers 9 and 6 both have a common factor of 3), then we would simplify the fraction by dividing both numbers by those factors until we get two numbers, p and q, which have no common factors. (In the above example, this will turn $\%$ into $\%$.) Clearly, p and q cannot both be even. (If they were, they would contain a common factor 2.) Our assumption is therefore that $p/q = \sqrt{2}$, where p and q are whole numbers that have no common factors. We can take the square of both sides to obtain: $p^2/q^2 = 2$. We will now multiply both sides by q^2 to obtain: $p^2 = 2q^2$. Notice that the right-hand side of this equation is clearly an even number, since it is some number (q^2) multiplied by 2, which always gives an even number. Since p^2 is equal to this even number, p^2 is an even number. But if the square of a number is even, then the number itself also must be even. (The square is simply the number multiplied by itself, and if the number were odd, its multiplication by itself would also be odd.) We therefore find that p itself must be an even number. Recall that this means that q must be odd, because p and q had no common factors. If, however, p is even, then we can write p in the form $p = 2r$ (since any even number has 2 as a factor). The previous equation $p^2 = 2q^2$ can therefore be written as (simply substituting $2r$ for p): $(2r)^2 = 2q^2$, which is [since $(2r)^2 = (2r) \times (2r)$] $4r^2 = 2q^2$. Dividing both sides by 2 gives: $2r^2 = q^2$. By the same arguments we used before, this says that q^2 is even (since it is equal to 2 times another number) and therefore that q *must be even.* Note, however, that above we showed that q must be odd! Thus we have reached something that is clearly a logical contradiction, since we showed that q must be odd and even at the same time. This fact demonstrates that our initial assumption, that there exist two whole numbers, p and q, the ratio of which is equal to $\sqrt{2}$, is false, thus completing the proof. Numbers like $\sqrt{2}$ represent a new kind of number—*irrational* numbers.

We can prove in a very similar way that the square root of any number that is not a perfect square (such as 9 or 16) is an irrational number. Numbers like $\sqrt{3}$, $\sqrt{5}$ are irrational numbers.

The magnitude of the discovery of incommensurability and irrational numbers cannot be overemphasized. Before this discovery, mathematicians had assumed that if you have any two line segments, one of which is longer than the other, then you can always find some smaller unit of measure so that the lengths of both segments will be exact whole-number multiples of this smaller unit. For example, if one segment is precisely 21.37 inches long and the other is 11.475 inches long, then we can measure both of them in units of one thousandth of an inch, and the first one will be 21,370 such units and the second 11,475 units. Early scholars therefore believed that finding such a common smaller measure was merely a matter of patient search. The discovery of incommensurability means that for the two segments of a line cut in a Golden Ratio (such as *AC* and *CB* in Figure 2), or for the diagonal and the side of a square, or for the diagonal and side of the pentagon, a common measure is never to be found. Steven Cushing published in 1988 a short poem (in the *Mathematics Magazine*) that describes our natural reaction to irrationals:

> *Pythagoras*
> *Did stagger us*
> *And our reason encumber*
> *With irrational number.*

We can appreciate better the intellectual leap that was required for the discovery of irrational numbers by realizing that even fractions, or *rational numbers* such as ½, ⅗, ¹¹⁄₁₃, represent by themselves an extremely important human discovery (or invention). The nineteenth-century mathematician Leopold Kronecker (1823–1891) expressed his opinion on this matter by saying: "God created the natural numbers, all else is the work of man."

Much of our knowledge about the familiarity of the ancient Egyptians with fractions, for example, comes from the Rhind (or Ahmes) Papyrus. This is a huge (about 18 feet long and 12 inches high) papyrus that was copied around 1650 B.C. from earlier documents by a scribe named Ahmes. The papyrus was found at Thebes and bought in 1858 by the Scottish antiquary Henry Rhind, and it is currently in the British Museum (except for a few fragments, which turned up unexpectedly in a collection of medical papers, and which are currently in the Brooklyn Museum). The Rhind Papyrus, which is in effect a calculator's handbook, has simple names only for unit fractions, such as ½, ⅓, ¼, etc., and for ⅔. A few other papyri have a name also for ¾. The ancient Egyptians generated other fractions simply by adding a few unit fractions. For example, they had ½ + ⅕ + ¹⁄₁₀ to represent ⅘ and ¹⁄₂₄ + ¹⁄₅₈ + ¹⁄₁₇₄ + ¹⁄₂₃₂ to represent ²⁄₂₉. To measure fractions of a capacity of grain called hekat, the ancient Egyptians used what were known as "Horus-eye" fractions. According to legend, in a fight between the god Horus, the son of Osiris and Isis, and the killer of his father, Horus' eye got torn away and broke into pieces. The god of writing and of calculations, Thoth, later found the pieces and wanted to restore the eye. However, he found only pieces that corresponded to the fractions ½, ¼, ⅛, ¹⁄₁₆, ¹⁄₃₂, and ¹⁄₆₄. Realizing that these fractions only add up to ⁶³⁄₆₄, Thoth produced the missing fraction of ¹⁄₆₄ by magic, which allowed him to complete the eye.

Strangely enough, the Egyptian system of unit fractions continued to be used in Europe for many centuries. For those during the Renaissance who had trouble memorizing how to add or subtract fractions, some writers of mathematical textbooks provided rules written in verse. An amusing example is provided by Thomas Hylles's *The Art of Vulgar Arithmetic, both in Integers and Fractions* (published in 1600):

Addition of fractions and likewise subtraction
Requireth that first they all have like bases
Which by reduction is brought to perfection
And being once done as ought in like cases,
Then add or subtract their tops and no more
Subscribing the base made common before.

In spite of, and perhaps (to some extent) because of, the secrecy surrounding Pythagoras and the Pythagorean Brotherhood, they are tentatively credited with some remarkable mathematical discoveries that may include the Golden Ratio and incommensurability. Given, however, the enormous prestige and successes of ancient Babylonian and Egyptian mathematics, and the fact that Pythagoras himself probably learned some of his mathematics in Egypt and Babylon, we may ask: Is it possible that these civilizations or others discovered the Golden Ratio even before the Pythagoreans? This question becomes particularly intriguing when we realize that the literature is bursting with claims that the Golden Ratio can be found in the dimensions of the Great Pyramid of Khufu at Giza. To answer this question, we will have to mount an exploratory expedition in archaeological mathematics.

3

UNDER A STAR-Y-POINTING PYRAMID?

The Pyramids first, which in Egypt were laid;
Next Babylon's Gardens, for Amytis made;
Then Mausolos' Tomb of affection and guilt;
Fourth, the Temple of Diana in Ephesus built;
The Colossus of Rhodes, cast in brass, to the Sun;
Sixth, Jupiter's Statue, by Phidias done;
The Pharos of Egypt comes last, we are told,
Or the Palace of Cyrus, cemented with gold.
—ANONYMOUS, SEVEN WONDERS
OF THE ANCIENT WORLD

The title of this chapter comes from the poem *On Shakespeare*, written in 1630 by the famous English poet John Milton (1608–1674). Milton, who himself was widely esteemed as a poet second only to Shakespeare, writes:

What needs my Shakespeare for his honor'd bones
The labor of an age in piled stones?
Or that his hallow'd relics should be hid
Under a star-y-pointing pyramid?
Dear son of memory, great heir of fame,
What need'st thou such weak witness of thy name?

As we shall soon see, the alignment of the pyramids was indeed based on the stars. As if these monuments are not awe-inspiring enough, however, many authors insist that the Great Pyramid's dimensions are based on the Golden Ratio. To all Golden Ratio enthusiasts, this association only adds to the air of mystique surrounding this proportion. But is this true? Did the ancient Egyptians really know about ϕ, and if they did, did they truly choose to "immortalize" the Golden Ratio by incorporating it into one of the Seven Wonders of the Ancient World?

Seeing that the initial interest in the Golden Ratio probably was inspired by its relation to the pentagram, we must first follow some of the early history of the pentagram, since this may lead us to the earliest occurrences of the Golden Ratio.

Ask any child to draw you a star and she will most likely draw a pentagram. This is actually a consequence of the fact that we happen to view stars through Earth's atmosphere. The turbulence of the air bends starlight in constantly shifting patterns, thus causing the familiar twinkling. In an attempt to represent the spikes generated by twinkling using a simple geometrical shape, humans came up with the pentagram, which also has the additional attractive property that it can be drawn without lifting the writing tool off the clay, papyrus, or paper.

Over the ages, such "stars" have become a symbol of excellence (e.g., five-star hotels, movies, book reviews), achievement ("stardom"), opportunity ("reach for the stars"), and authority ("five-star" generals). When this symbolism is combined with the romantic appeal of a starry night, it is no wonder that the flags of more than sixty nations depict five-pointed stars and that such star patterns appear on innumerable commercial logos (e.g., Texaco, Chrysler).

Some of the earliest known pentagrams come from fourth millennium B.C. Mesopotamia. Pentagram shapes were found in excavations in Uruk (where the earliest writings were also uncovered), and in Jemdet Nasr. The ancient Sumerian city of Uruk is probably also the one mentioned in the Bible (Genesis 10) as Erech, one of the cities in the kingdom of the mighty hunter Nimrod. The pentagram was found on a clay tablet dated to about 3200 B.C. In Jemdet Nasr, pentagrams from about the same period were found on a vase and on a spindle whorl. In Sumerian the pentagram, or its cuneiform derivative, meant "the regions of the universe." Other parts of the ancient Middle East also produced pentagrams. A pentagram on a flint scraper from the Chalcolithic period (4500–3100 B.C.) was found at Tel Eshdar in the Israeli Negev Desert. Pentagrams were also found in Israel in excavations at Gezer and at Tel Zachariah, but those date to a considerably later period (the fifth century B.C.). In spite of the fact that five-pointed stars appear quite frequently in ancient Egyptian artifacts, true geometrical pentagrams are not very common, although a pentagram dating to around 3100 B.C. was found on a jar in Naqadah, near Thebes. Generally, the hieroglyphic symbol of a star enclosed in a circle meant the "underworld," or the mythical dwelling of stars at twilight time, while stars without circles served simply to signify the night stars.

The main question we need to answer, however, in the context of this book is not whether pentagrams or pentagons had any symbolic or mystic meanings for these early civilizations but whether these civilizations were also aware of the *geometrical* properties of these figures and, in particular, of the Golden Ratio.

ERE BABYLON WAS DUST

Studies of cuneiform tablets dating to the second millennium B.C., which were discovered in 1936 in Susa in Iran, leave very little doubt that the Babylonians of the first dynasty knew at least an approximate formula for the area of a pentagon. The Babylonian interest in the pentagon may have originated from the simple fact that this is the figure obtained if the tips of all five fingers are pressed against a clay tablet. A line on a Susa tablet reads: "1 40, the constant of the five-sided figure." Since the Babylonians used the sexagesimal (base 60) system, the numbers 1 40 should be interpreted at $1 + 40/60$, or $1.666 \ldots$, for the area of the pentagon. The actual area of a pentagon with a side of unit length is, in fact, not too far from this value—1.720. The Babylonians had a similar approximation for pi, the ratio of the circumference of a circle to its diameter. In fact, the approximations for both pi and the area of the pentagon relied on the same relation. The Babylonians assumed that the perimeter of any regular polygon (shape of many equal sides and angles) is equal to six times the radius of the circle circumscribing that polygon (Figure 12). This relation is actually precise for a regular hexagon (six-sided shape; Figure 12), since all the triangles have equal sides. The value of π deduced by the Babylonians was $\pi = 3\frac{1}{8} = 3.125$. This is really not a bad approximation, given that the precise value is 3.14159 . . . For the pentagon, the assumption "perimeter equals six times the radius" (which is not accurate) gives the approximate value 1.666 . . . for the area that appears in the Susa tablet.

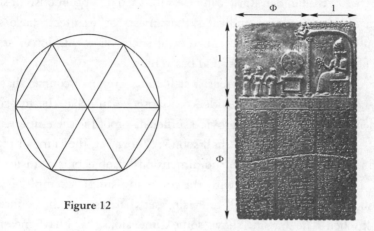

Figure 12

Figure 13

In spite of these significant early discoveries in mathematics and the intimate relation between the pentagon-pentagram system and the Golden Ratio, there is absolutely no shred of mathematical evidence that the Babylonians knew about the Golden Ratio. Nevertheless, some texts claim that the Golden Ratio is found on Babylonian and Assyrian stelae and bas-reliefs. For example, a Babylonian stela (Figure 13) depicting priests leading an initiate to a "meeting" with the sun god is said (in Michael Schneider's entertaining book, *A Beginner's Guide to Constructing the Universe*) to contain "many Golden Ratio relationships." Similarly, in an article that appeared in 1976 in the journal *The Fibonacci Quarterly*, art analyst Helene Hedian states that a bas-relief of an Assyrian winged demigod of the ninth century B.C. (currently in the Metropolitan Museum of Art) fits perfectly into a rectangle with dimensions that are in a Golden Ratio. Furthermore, Hedian suggests that the strong lines of the wings, legs, and beak follow other phi divisions. Hedian also makes a similar assertion about the Babylonian "Dying Lioness" from Nineveh, dated to around 600 B.C., which is currently in the British Museum in London.

Does the Golden Ratio really feature in these Mesopotamian artifacts, or is this merely a misconception?

In order to answer this question, we must be able somehow to identify criteria that will allow us to determine whether certain claims about the appearance of the Golden Ratio are true or false. Clearly, the presence of the Golden Ratio can be established unambiguously if some form of documentation indicates that artists or architects have consciously made use of it. Unfortunately, no such documentation exists for any of the Babylonian tablets and bas-reliefs.

A devoted Golden Numberist still could argue, of course, that the absence of evidence is not evidence of absence and that the measured dimensions by themselves provide sufficient proof for the employment of the Golden Ratio. As we shall soon see, however, the game of trying to find the Golden Ratio in the dimensions of objects is a misleading one. Let me illustrate this with the following simple example. Figure 14 shows a sketch of a small television set that rests on the counter in my kitchen. The drawing shows some dimensions that I have measured

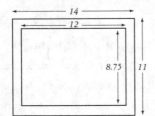

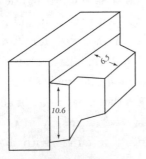

by myself. You will notice that the ratio of the height of the protrusion at the television's rear to its width, 10.6/6.5 = 1.63, and the ratio of the length of the front to the height of the screen, 14/8.75 = 1.6, are both in reasonable agreement with the value of the Golden Ratio, 1.618 . . . . Does this mean that the makers of this television decided to include the Golden Ratio in its architecture? Clearly not. This example simply demonstrates the two main shortcomings of claims about the presence of the Golden Ratio in architecture or in works of art, on the basis of dimensions alone: (1) they involve numerical juggling, and (2) they overlook inaccuracies in measurements.

Figure 14

Any time you measure the dimensions of some relatively complicated structure (e.g., a picture on a stela or a television set), you will
have at your disposal an entire collection of lengths to choose from. As
long as you can conveniently ignore parts of the object under consideration, if you have the patience to juggle and manipulate the numbers in
various ways, you are bound to come up with some interesting numbers. Thus, in the television, I was able to "discover" some dimensions
that give ratios that are close to the Golden Ratio.

The second point that is often ignored by the too-passionate
Golden Ratio aficionados is that any measurements of lengths involve
errors or inaccuracies. It is important to realize that any inaccuracy in
length measurements leads to a yet larger inaccuracy in the calculated
ratio. For example, imagine that two lengths, of 10 inches each, are
measured with a precision of 1 percent. This means that the result of the
measurement for each length could be anywhere between 9.9 and 10.1
inches. The ratio of these measured lengths could be as bad as 9.9/10.1
= 0.98, which represents a 2 percent inaccuracy—double that of the individual measurements. Therefore, an overzealous Golden Numberist
could change two measurements by only 1 percent, thereby affecting
the obtained ratio by 2 percent.

A reexamination of Figure 13 with these caveats in mind reveals,
for example, that the vertical long segment has been conveniently chosen so as to include the base of the bas-relief and not just the cuneiform
text. Similarly, the point to which the horizontal long segment was
measured has been chosen quite arbitrarily to be to the right, rather
than to the left, of the edge of the bas-relief.

Reevaluating all of the existing material in this light, I have to conclude that it is very unlikely that the Babylonians discovered the
Golden Ratio.

WAY DOWN IN EGYPT LAND

The situation concerning the ancient Egyptians is more complicated,
and it requires a considerable amount of detective work. Here we are
confronted with what is suggested to be overwhelming evidence in the

form of numerous texts that claim that phi can be found, for example, in the proportions of the Great Pyramid and in other ancient Egyptian monuments.

Let me start with two of the easier cases, those of the Osirion and the Tomb of Petosiris. The Osirion is a temple considered to be the cenotaph of King Seti I, who ruled Egypt in the XIX dynasty from about 1300 to about 1290 B.C. The temple was discovered by the noted archaeologist Sir Flinders Petrie in 1901, and the massive excavation works were completed in 1927. The temple itself is supposed to represent, via its architectural symbolism, the myth of Osiris. Osiris, the husband of Isis, was originally the king of Egypt. His brother Seth murdered him and scattered the pieces of his body. Isis collected the pieces, thus providing Osiris with a renewed life. Consequently, Osiris became king of the Underworld and of cyclic transformation through death and rebirth on both the individual and cosmic levels. After the cult of the dead was further developed during the Middle Kingdom (2000–1786 B.C.), Osiris was regarded as the judge of the soul after death.

The entire roofed Osirion temple was covered with earth, so as to resemble an underground tomb. The plan of the Osirion (Figure 15a)

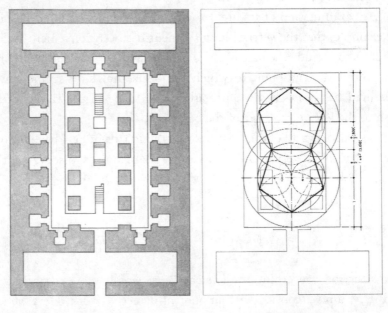

Figure 15a   Figure 15b

contains a central area with ten square columns, which is surrounded by what was probably a water-filled ditch. This structure has been interpreted to symbolize creation out of the primordial waters.

In his interesting 1982 book *Sacred Geometry: Philosophy and Practice*, Robert Lawlor suggests that the geometry of the Osirion is "conforming to the proportions of the Golden Section" because "the Golden Proportion is the transcendent 'idea-form' which must exist *a priori* and eternally before all the progressions which evolve in time and space." To support his suggestion about the prominent appearance of the Golden Ratio, ϕ, in the architectural design of the temple, Lawlor offers detailed geometrical analyses of the type presented in Figure 15b. Furthermore, he claims that "the emphasis on the theme of the pentagon aptly symbolized the belief that the king, after death, became a star."

In spite of their considerable visual appeal, I find Lawlor's analyses unconvincing. Not only are the lines that are supposed to indicate Golden Ratio proportions drawn at what appear to be totally arbitrary points, but even the pentagons represent, in my opinion, a rather forced interpretation of what is basically a rectangular shape. The fact that Lawlor himself presents other geometrical analyses of the temple's geometry (with ϕ being associated with different dimensions) further demonstrates the nonunique and somewhat capricious nature of such readings.

The situation with the Tomb of Petosiris, which was excavated by archaeologist Gustave Lefebvre during the early 1920s, is very similar. The tomb is not as old as the Osirion, dating only to about 300 B.C., and it was built for the High Priest (known as Master of the Seat) of Thoth. Since this tomb is from a period during which the Golden Ratio was already known (to the Greeks), at least in principle, the Golden Ratio could feature in the tomb's geometry. In fact, Lawlor (again in *Sacred Geometry*) concludes that "the Master Petosiris had a complete and extremely sophisticated knowledge of the Golden Proportion." This conclusion is based on two geometrical analyses of a painted bas-relief from the east wall of the tomb's chapel (Figure 16a). The bas-relief shows a priest pouring a libation over the head of the mummy of the deceased.

Unfortunately, the geometrical analyses that Lawlor presents appear

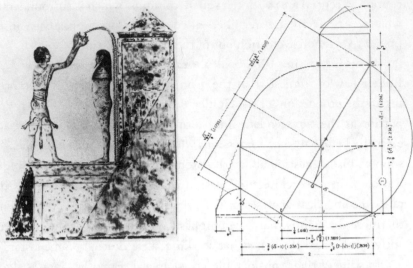

Figure 16a   Figure 16b

rather contrived (Figure 16b), with lines drawn conveniently at points that are not obvious terminals at all. Furthermore, some of the ratios obtained are too convoluted (e.g., $2\sqrt{1 + \Phi^2} / \Phi^2$) to be credible. My personal feeling is, therefore, that while Lawlor's assertion that "the burial practices in the Pharaonic tradition were undertaken not merely to provide a receptacle for the physical body of the deceased, but also to make a place to retain the metaphysical knowledge which the person had mastered in his lifetime" is a very correct one, the Golden Ratio is unlikely to have been a part of Petosiris' knowledge.

I should emphasize that it is virtually impossible to prove that the Golden Ratio does not appear in some Egyptian artifacts when the evidence is presented only in the form of some measured dimensions. However, in the absence of any supporting documentation, the dimensions of the artwork or architectural design have to be such that the Golden Ratio will literally jump at you, rather than be buried so deeply to require a very complex analysis to be revealed. As we shall see later, detailed investigations of several much more recent cases for which claims existed in the literature that the artists had used the Golden Ratio show that there was no basis for these assertions.

Instead of continuing with relatively obscure objects, such as an

Egyptian stela from around 2150 B.C., claimed by some to show dimensions in a ratio of φ, let me now turn to the main event—the Great Pyramid of Khufu.

PYRAMID OF NUMBERS

According to tradition, it was King Menes (or Narmer) who as a ruler of Upper Egypt conquered the rival kingdom of Lower Egypt (in the delta of the Nile), thus uniting Egypt as a single kingdom, around 3110 B.C. Sun worship as the basic form of religion was introduced under the rule of the third dynasty (ca. 2780–2680 B.C.), as were mummification and the construction of large stone monuments. The age of the great pyramids reached its climax during the fourth dynasty, around 2500 B.C., in the famous triad of pyramids at Giza (Figure 17). The "Great Pyramid" (the one at the back in the figure) stands not only as a monument to the king but also as a testimony to the success of a unified organization of the ancient Egyptian society. Researcher Kurt Mendelssohn concluded in his 1974 book *The Riddle of the Pyramids* that, to

Figure 17

a large extent, the object of the whole exercise of constructing the pyramids was not the use to which the final products were to be put (to serve as tombs), but their manufacture. In other words, what mattered was not the pyramids themselves but the *building* of the pyramids. This would explain the apparent disparity between the tremendous effort of piling up some 20 million tons of quarried limestone and the sole purpose of burying under them three pharaohs.

In 1996 amateur Egyptologist Stuart Kirkland Wier, working under the sponsorship of the Denver Museum of Natural History, estimated that building the Great Pyramid at Giza required something like 10,000 workers. A calculation of the energy required to carry the blocks of stone from the quarry to the pyramid site, as well as that needed to lift the stones to the necessary height, gave Wier the total amount of work that had to be invested. Assuming that the construction lasted twenty-three years (the length of King Khufu's reign), and making some reasonable assumptions about the daily energy output of an Egyptian worker and about the construction schedule, Wier was able to estimate the size of the workforce.

Until recently, the dating of the pyramids at Giza relied mostly on surviving lists of kings and the lengths of their reigns. Since these lists are rare, seldom complete, and known to contain inconsistencies, chronologies generally were accurate only to within about a hundred years. (Dating by radioactive carbon contains a similar uncertainty.) In a paper that appeared in the journal *Nature* in November 2000, Kate Spence of Cambridge University proposed another method of dating, which gives for Khufu's Great Pyramid a date of 2480 B.C., with an uncertainty of only about five years. Spence's method is the one originally suggested by the astronomer Sir John Herschel in the middle of the nineteenth century, and it is based on the fact that the pyramids were always oriented with respect to the north direction with extraordinary precision. For example, the orientation of the Great Pyramid at Giza deviates from the true north by less than 3 minutes of arc (a mere 5 percent of one degree). There is no doubt that the Egyptians used astronomical observations to determine the north direction with such accuracy.

The north celestial pole is defined as a point on the sky aligned with
Earth's rotation axis, around which the stars appear to rotate. However,
the axis of Earth itself is not precisely fixed; rather it wobbles very
slowly like the axis of a spinning top or gyroscope. As a result of this
motion, known as precession, the north celestial pole appears to trace
out a large circle on the northern sky about every 26,000 years. While
today the north celestial pole is marked (to within one degree) by the
North Star, Polaris (known by the astronomical name of α-Ursae Minoris), this was not the case at the time of the Great Pyramid's construction. By tentatively identifying the two stars that the ancient
Egyptians used to mark the north to be ζ-Ursae Majoris and β-Ursae
Minoris, and by a careful examination of the alignments of eight pyramids, Spence was able to determine the date of accession of Khufu's
pyramid to be 2480 B.C. ± 5, about seventy-four years younger than
previous estimates.

Few archaeological structures have generated as much myth and
controversy as has the Great Pyramid. The preoccupation with the pyramid, or the occult side of pyramidology, was, for example, a central
theme to the cult of the Rosicrucians (founded in 1459 by Christian
Rosenkreuz). The members of this cult made great pretensions to
knowledge of the secrets of nature, magical signatures, and the like.
Freemasonry originated from some factions of the Rosicrucians' cult.

The more modern interest in pyramidology started probably with
the religiously permeated book of the retired English publisher John
Taylor, *The Great Pyramid: Why Was It Built and Who Built It?* which
appeared in 1859. Taylor was so convinced that the pyramid contained
a variety of dimensions inspired by mathematical truths unknown to
the ancient Egyptians that he concluded that its construction was the
result of divine intervention. Influenced by the then-fashionable idea
that the British were the descendants of the lost tribes of Israel, he proposed, for example, that the basic measuring unit of the pyramid was
the same as the biblical cubit (slightly more than 25 British inches;
equal to precisely 25 "pyramid inches"). This unit supposedly was also
the one employed by Noah in building the Ark and by King Solomon
in the construction of the Temple. Taylor went on to claim that this

sacred cubit was divinely selected on the basis of the length of Earth's center-to-pole radius, with the "pyramid inch" being the five-hundred-millionth part of Earth's polar axis. His cranky book found a great admirer in Charles Piazzi Smyth, the Astronomer-Royal of Scotland, who published no fewer than three massive tomes (the first entitled *Our Inheritance in the Great Pyramid*) on the Great Pyramid in the 1860s. Piazzi Smyth's enthusiasm was motivated partly by his strong objection to attempts to introduce the metric system in Britain. His pseudoscientific/theological logic worked something like this: The Great Pyramid was designed in inches; the mathematical properties of the pyramid show that it was constructed by divine inspiration; therefore, the inch is a God-given unit, unlike the centimeter, which was inspired "by the wildest, most blood-thirsty and most atheistic revolution" (meaning the French Revolution). In further describing his views on the system of measures debate, Piazzi Smyth writes (in *The Great Pyramid, Its Secrets and Mysteries Revealed*):

> So that not for the force of the sparse oratory emitted in defense of British metrology before Parliament, were the bills of the pro-French metrical agitators so often overthrown, but for the sins rather of that high-vaulting system itself; and to prevent a chosen nation, a nation preserved through history . . . to prevent that nation unheedingly robing itself in the accursed thing, in the very garment of the coming Anti-Christ; and Esau-like, for a little base-pottage, for a little temporarily extra mercantile profit, throwing away a birthright institution which our Abrahamic race was intended to keep, until the accomplishment of the mystery of God touching all humankind.

After reading this text, we cannot be too surprised to find out that author Leonard Cottrell chose to entitle the chapter on Charles Piazzi Smyth in his book *The Mountains of Pharaoh* "The Great Pyramidiot."

Both Piazzi Smyth and Taylor essentially revived the Pythagorean obsession with the number 5 in their numerology-based analysis of the pyramid. They noted that the pyramid has, of course, five corners and five faces (counting the base); that the "sacred cubit" had about 25 (5 squared) inches (or precisely 25 "pyramid inches"); that the "pyramid inch" is the five-hundred-millionth part of Earth's axis; and so on.

Writer Martin Gardner found an amusing example that demonstrates the absurdity in Piazzi Smyth's "fiveness" analysis. In his book *Fads and Fallacies in the Name of Science,* Gardner writes:

> If one looks up the facts about the Washington Monument in the *World Almanac,* he will find considerable fiveness. Its height is 555 feet and 5 inches. The base is 55 feet square, and the windows are set at 500 feet from the base. If the base is multiplied by 60 (or five times the number of months in a year) it gives 3,300, which is the exact weight of the capstone in pounds. Also, the word "Washington" has exactly ten letters (two times five). And if the weight of the capstone is multiplied by the base, the result is 181,500—a fairly close approximation to the speed of light in miles per second.

Here, however, comes the most dramatic announcement concerning the Great Pyramid in the context of our interest in the Golden Ratio. In the same book, Gardner refers to a statement that, if true, shows that the Golden Ratio was actually incorporated in the Great Pyramid by design. Gardner writes: "Herodotus states that the Pyramid was built so the area of each face would equal the area of a square whose side is equal to the Pyramid's height." The Greek historian Herodotus (ca. 485–425 B.C.) was called "the Father of History" by the great Roman orator Cicero (106–43 B.C.). While Gardner did not realize the full implications of Herodotus' statement, he was neither the first nor the last to present it.

In an article entitled "British Modular Standard of Length," which appeared in *The Athenaum* on April 28, 1860, the famous British astronomer Sir John (Frederick William) Herschel (1792–1871) writes:

> The same slope, . . . belongs to a pyramid characterized by the property of having each of its faces equal to the square described upon its height. This is the characteristic relation which, Herodotus distinctly tells us, it was the intention of its builders that it should embody, and which we now know that it did embody.

Most recently, in 1999, French author and telecommunications expert Midhat J. Gazalé writes in his interesting book *Gnomon: From Pharaohs to Fractals:* "It was reported that the Greek historian Herodotus learned from the Egyptian priests that the square of the Great Pyramid's height is equal to the area of its triangular lateral side." Why is this statement so crucial? For the simple reason that it is equivalent to saying that the Great Pyramid was designed so that the ratio of the height of its triangular face to half the side of the base is equal to the Golden Ratio!

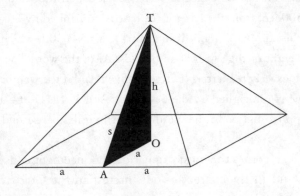

Figure 18

Examine for a moment the sketch of the pyramid in Figure 18, in which *a* is half the side of the base, *s* is the height of the triangular face, and *h* is the pyramid's height. If the statement attributed to Herodotus is correct, this would mean that h^2 (the square of the pyramid's height) is equal to $s \times a$ (the area of the triangular face; see Appendix 3). Some elementary geometry shows that this equality means that the ratio s/a is precisely equal to the Golden Ratio. (The proof is given in Appendix 3.) The immediate question that comes to mind is: Well, is it? The base of the Great Pyramid is actually not a perfect square, as the lengths of the sides vary from 755.43 feet to 756.08 feet. The average of the lengths is $2a = 755.79$ feet. The height of the pyramid is $h = 481.4$ feet. From these values we find (by using the Pythagorean theorem) that the height of the triangular side *s* is equal to $s = 612.01$ feet. We therefore find that $s/a = 612.01/377.90 = 1.62$, which is indeed extremely close to (differing by less than 0.1 percent from) the Golden Ratio.

Taken at face value, therefore, this evidence would imply that the ancient Egyptians indeed knew about the Golden Ratio, since not only does this number appear in the ratio of dimensions of the Great Pyramid but its presence seems to be supported by historical documentation of the intentions of the designers, in the form of Herodotus' statement. But is this true? Or are we witnessing here what the Canadian mathematician and author Roger Herz-Fischler called "one of the most ingenious sleights of hand in 'scientific' history"?

Clearly, since the measurements of the dimensions cannot be altered, the only part in this "evidence" for the presence of the Golden Ratio that can be challenged is Herodotus' statement. In spite of the numerous repetitions of the quote from *History,* and even though one cannot cross-examine a man who lived 2,500 years ago, at least four researchers have taken upon themselves the "detective" work of investigating what Herodotus really said or meant. The results of two of these inquiries have been summarized by Herz-Fischler and by University of Maine mathematician George Markowsky.

The original text from Herodotus' *History* appears in paragraph 124 of book II, named *Euterpe.* Traditional translations read: "Its base is square, each side is eight plethra long and its height the same," or "It is a square, eight hundred feet each way, and the height the same." Note that one plethron was 100 Greek feet (approximately 101 English feet).

These texts look very different from what has been presented as a quote (that the square of the height equals the area of the face) from Herodotus. Furthermore, the figures for the pyramid's dimensions that Herodotus mentions are wildly off. The Great Pyramid is far from being 800 feet high (its height is only about 481 feet), and even the side of its square base (about 756 feet) is significantly less than 800 feet. So where did that "quote" come from? The first clue comes from Sir John Herschel's article in *The Athenaeum.* According to Herschel, it was John Taylor in his book *The Great Pyramid, Why Was It Built and Who Built It?* who had "the merit of pointing out" this property of the pyramid and Herodotus' quote. Herz-Fischler tracked down the misconception to what appears to be nothing more than a misinterpretation of Herodotus in John Taylor's by now infamous book.

Taylor starts with a translation of Herodotus that does not read too differently from the ones above: "of this Pyramid, which is four-sided,

each face is, on every side 8 plethra, and the height equal." Here, however, he lets his imagination run wild, by assuming that Herodotus
meant that the number of square feet in each face is equal to the number of square feet in a square with a side equal to the pyramid's height.
Even with this "imaginative" interpretation, Taylor is still left with the
small problem that the number mentioned (eight plethra) is way off the
actual measurements. His suggested solution to this problem is even
more appalling. With no justification whatsoever, he claims that the
eight plethra must be multiplied by the area of the base of one of the
small pyramids standing on the east side of the Great Pyramid.

The conclusion from all of this is that Herodotus' text can hardly be
taken as documenting the presence of the Golden Ratio in the Great
Pyramid. The totally unfounded interpretation of the text instigated by
Taylor's book (and subsequently repeated endless times) really makes
little sense and represents just another case of number juggling.

Not everyone agrees with this conclusion. In an article entitled
"The Icosahedral Design of the Great Pyramid," which appeared in
1992, Hugo F. Verheyen proposes that the Golden Ratio as a mystic
symbol may have been deliberately hidden within the design of the
Great Pyramid "as a message for those who understand." As we shall see
later, however, there are more reasons to doubt the idea that the Golden
Ratio featured at all in the pyramid's design.

When we realize that the Great Pyramid rivals the legendary city of
Atlantis in the numbers of books written about it, we should not be too
surprised to hear that ϕ was not the only special number to be invoked
in pyramidology—π was too.

The π theory appeared first in 1838, in *Letter from Alexandria, on the
Evidence of the Practical Application of the Quadrature of the Circle, in the
Configuration of the Great Pyramids of Egypt,* by H. Agnew, but it is generally credited to Taylor, who merely repeated Agnew's theory. The
claim is that that ratio of the circumference of the base ($8a$ in our previous notation, in which a was half the side of the base) to the pyramid's
height (h) is equal to 2π. If we use the same measured dimensions we
used before, we find $8a/h = 4 \times 755.79/481.4 = 6.28$, which is equal to
2π with a remarkable precision (differing only by about 0.05 of a percent).

The first thing to note, therefore, is that just from the dimensions of the Great Pyramid alone, it would be impossible to determine whether phi or pi, *if either,* was a factor in the pyramid's design. In fact, in an article published in 1968 in the journal *The Fibonacci Quarterly,* Colonel R. S. Beard of Berkeley, California, concluded that: "So roll the dice and choose your own theory."

If we have to choose between π and ϕ as potential contributors to the pyramid's architecture, then π has a clear advantage over ϕ. First, the Rhind (Ahmes) Papyrus, one of our main sources of knowledge of Egyptian mathematics, informs us that the ancient Egyptians of the seventeenth century B.C. knew at least an approximate value of π, while there is absolutely no evidence that they knew about ϕ. Recall that Ahmes copied this mathematical handbook at about 1650 B.C., during the period of the Hyksos or shepherd kings. However, he references the original document from the time of King Ammenemēs III of the Twelfth Dynasty; and it is perhaps not impossible (although it is unlikely) that the contents of the document had already been known at the time of the construction of the Great Pyramid. The papyrus contains eighty-seven mathematical problems preceded by a table of fractions. There is considerable evidence (in the form of other papyri and records) that the table continued to serve as a reference for nearly two thousand years. In his introduction, Ahmes describes the document as "the entrance into the knowledge of all existing things and all obscure secrets." The Egyptian estimate of π appears in problem number 50 of the Rhind Papyrus, which deals with determining the area of a circular field. Ahmes' solution suggests: "take away ⅑ of the diameter and square the remainder." From this we deduce that the Egyptians approximated π to be equal to $^{256}/_{81} = 3.16049\ldots$, which is less than 1 percent off the correct value of $3.14159\ldots$.

A second fact that gives π an advantage over ϕ is the interesting theory that the builders incorporated π into the pyramid's design even without knowing its value. This theory was put forward by Kurt Mendelssohn in *The Riddle of the Pyramids.* Mendelssohn's logic works as follows. Since there is absolutely no evidence that the Egyptians at the time of the Old Kingdom had anything but the most rudimentary command of mathematics, the presence of π in the pyramid's geometry must be the consequence of some practical, rather than theoretical, design concept. Mendelssohn suggests that the ancient Egyptians may have not used the same unit of length to measure vertical and horizontal distances. Rather, they could have used palm fiber ropes to measure the height of the pyramid (in units of cubits) and roller drums (one cubit in diameter) to measure the length of the base of the pyramid. In this way, horizontal lengths would have been obtained by counting the revolutions in units one might call "rolled cubits." All the Egyptian architect then had to do was to choose how many cubits he wanted his workers to build upward for every horizontal rolled cubit. Since one rolled cubit is really equal to π cubits (the circumference of a circle with a diameter of one cubit), this method of construction would imprint the value of π into the pyramid's design without the builders even knowing it.

Of course, there is no way to test Mendelssohn's speculation directly. However, some Egyptologists claim that there does exist direct evidence suggesting *that neither the Golden Ratio nor pi* were used in the Great Pyramid's design (not even inadvertently). This theory is based on the concept of the *seked.* The seked was simply a measure of the slope of the sides of a pyramid or, more precisely, the number of horizontal cubits needed to move for each vertical cubit. Clearly, this was an important practical concept for the builders, who needed to keep a constant shape with each subsequent block of stone. The problems numbered 56 to 60 in the Rhind Papyrus deal with calculations of the seked and are described in great detail in Richard J. Gillings's excellent book, *Mathematics in the Time of the Pharaohs.* In 1883, Sir Flinders Petrie found that the choice of a particular seked (slope of the pyramid's side) gives for the Great Pyramid the property of "ratio of circumference of the base to the pyramid's height equal to 2π" to a high precision, with π playing no role whatsoever in the design. Supporters of the seked hypothesis point out that precisely the same seked is found in the step pyramid at Meidum, which was built just before the Great Pyramid at Giza.

Not all agree with the seked theory. Kurt Mendelssohn writes: "A great number of mathematical explanations have been suggested and even one, made by a noted archaeologist [Petrie], that the builders by accident used a ratio of $^{14}\!/_{11}$ [$= {}^{28}\!/_{22}$, which is very close to $4/\pi$], remains lamentably unconvincing." On the other hand, Roger Herz-Fischler, who examined no fewer than nine theories that have been advanced for the Great Pyramid's design, concluded in a paper that appeared in the journal *Crux Mathematicorum* in 1978 that the seked theory is very probably the correct one.

From our perspective, however, if either of the two hypotheses, seked or rollers, is correct, then the Golden Ratio played no role in the Great Pyramid's design.

Is, therefore, the 4,500-year-old case of the Golden Ratio and the Great Pyramid closed? We would certainly hope so, but unfortunately history has shown that the mystical appeal of the pyramids and Golden Numberism may be stronger than any solid evidence. The arguments presented by Petrie, Gillings, Mendelssohn, and Herz-Fischler have been available for decades, yet this has not prevented the publication of numerous new books repeating the Golden Ratio fallacy.

For our purposes, we have to conclude that it is highly unlikely that either the ancient Babylonians or the ancient Egyptians discovered the Golden Ratio and its properties; this task was left for the Greek mathematicians.

4

THE SECOND
TREASURE

There is no doubt that anybody who grew up in a western or mideast-ern civilization is a pupil of the ancient Greeks, when it comes to math-ematics, science, philosophy, art, and literature. The phrase of the German poet Goethe—"of all peoples the Greeks have dreamt the dream of life the best"—is only a small tribute to the pioneering efforts of the Greeks in branches of knowledge that they invented and denom-inated.

However, even the accomplishments of the Greeks in many other fields pale in comparison with their awe-inspiring achievements in mathematics. In the span of only four hundred years, for example, from Thales of Miletus (at ca. 600 B.C.) to "the Great Geometer" Apollonius of Perga (at ca. 200 B.C.), the Greeks completed all the essentials of a theory of geometry.

The Greek excellence in mathematics was largely a direct consequence of their passion for knowledge for its own sake, rather than merely for practical purposes. A story has it that when a student who learned one geometrical proposition with Euclid asked, "But what do I gain from this?" Euclid told his slave to give the boy a coin, so that the student would see an actual profit.

The curriculum for the education of statesmen at the time of Plato included arithmetic, geometry, solid geometry, astronomy, and music—all of which, the Pythagorean Archytas tells us, fell under the general definition of "mathematics." According to legend, when Alexander the Great asked his teacher Menaechmus (who is reputed to have discovered the curves of the ellipse, the parabola, and the hyperbola) for a shortcut to geometry, he got the reply: "O King, for traveling over the country there are royal roads and roads for common citizens; but in geometry there is one road for all."

PLATO

Into this intellectual milieu enter Plato (428/427 B.C.–348/347 B.C.), one of the most influential minds of ancient Greece and of western civilization in general. Plato is said to have studied mathematics with the Pythagorean Theodorus of Cyrene, who was the first to prove that not just $\sqrt{2}$ but also numbers like $\sqrt{3}$, $\sqrt{5}$, and up to $\sqrt{17}$ are irrational. (No one knows for sure why he stopped at 17, but clearly he did not have a general proof.) Some researchers claim that Theodorus also may have used a line cut in the Golden Ratio to provide what may be the easiest proof of incommensurability. (The idea is essentially the same as that presented in Appendix 2.)

As Plato states in *The Republic,* mathematics was an absolute must in the education of all state leaders and philosophers. Accordingly, the inscription over the entrance to his school (the Academy) read: "Let no one destitute of geometry enter my doors." The historian of mathematics David Eugene Smith describes this in his book *Our Debt to Greece and Rome* as the first college entrance requirement in history. Plato's admiration for mathematics also shows when he speaks with some envy on the attitude toward mathematics in Egypt, where "arithmetical games have been invented for the use of mere children, which they learn as a pleasure and amusement."

In considering the role of Plato in mathematics in general, and in relation to the Golden Ratio in particular, we have to examine not just his own purely mathematical contributions, which were not very significant, but the effects of his influence and encouragement on the mathematics of others of his and of later generations. To some extent, Plato may be considered as one of the first true theoreticians. His theoretical inclinations are best exemplified by his attitude toward astronomy, where, rather than observing the stars in their motions, he advocates to "leave the heavens alone" and concentrate on the more abstract heaven of mathematics. The latter, according to Plato, is merely represented by the actual stars, in the same way that the abstract entities of a point, a line, and a circle are represented by geometrical drawings. Interestingly, in his outstanding book *A History of Greek Mathematics* (1921), Sir Thomas Heath writes: "It is difficult to see what Plato can mean by the contrast which he draws between the visible broideries of heaven [the visible stars and their arrangement], which are indeed beautiful, and the true broideries which they only imitate and which are infinitely more beautiful and marvelous."

As a theoretical astrophysicist myself, I must say that I feel quite a bit of affinity with some of the sentiments expressed by Plato's underlying motif. The distinction here is between the beauty of the cosmos itself and the beauty of the theory that explains the universe. Let me clarify this with a simple example, the principle of which was first discovered by the famous German painter Albrecht Dürer (1471–1528).

You can put together six pentagons (Figure 19) to make one large pentagon, with five holes in the shape of *Golden Triangles* (isosceles triangles with a ratio of side to base of ϕ). Six of these pentagons, in turn, go together to make an even larger (and more holey-looking) pentagon, and so on indefinitely.

I think everyone will agree that the obtained shape (Figure 19) is extremely beautiful. But this shape contains an additional mathematical appeal, which is in the simplicity of the underlying principle of its

Figure 19

construction. This is, I believe, that mathematical heaven to which Plato was referring.

There is little doubt that Plato's guidance was far more important than his direct contributions. A text attributed to Philodemus from the first century reads: "Great progress in mathematics [was achieved] during that time, with Plato as the director and problem-giver, and the mathematicians investigating them zealously."

Nevertheless, Plato himself certainly had an intense interest in the properties of numbers and of geometrical figures. In *Laws,* for example, he suggests that the optimal number of citizens in a state is 5,040 because: (a) it is the product of 12, 20, and 21; (b) the twelfth part of it can still be divided by 12; (c) it has 59 divisors, including all the whole numbers from 1 to 12 (except for 11, but 5,038, which is very close to 5,040, is divisible by 11). The choice of this number and its properties allow Plato to develop his socioeconomic vision. For example, the state's land is divided into 5,040 lots, with 420 such lots constituting the territory of each of twelve "tribes." The people of the state themselves are divided into four social categories: free citizens and their wives and children, their slaves, resident aliens, and a diverse population of visiting aliens. In elections to the Council, members from all four property categories vote for ninety members from each class.

Another number often associated with Plato is 216. Plato mentions this number in *The Republic* in a rather obscure passage that alludes to the fact that 216 is equal to 6 cubed, with 6 being one of the numbers representing marriage (since it is the product of the female 2 and the male 3). Plato, himself a pupil of the Pythagoreans, was also aware of the fact that the sum of the cubes of the sides of the famous 3-4-5 Pythagorean triangle is equal to 216.

Plato and the Golden Section are linked mainly through two areas that were particularly close to his heart: incommensurability and the *Platonic solids.* In *Laws,* Plato expresses his own feeling of shame for having learned about incommensurable lengths and irrational numbers relatively late in his life, and he laments the fact that many of the Greeks of his generation were still not familiar with the existence of such numbers.

Plato recognizes (in *Hippias Major*) that just as an even number may be the sum of either two even or two odd numbers, so can the sum of two irrationals be either irrational or rational. Since we already know that ϕ is irrational, a rational straight line (e.g., of unit length) divided in a Golden Section provides an illustration of the latter case, although Plato may not have known this fact. Some researchers maintain that Plato had a direct interest in the Golden Section. They point out that when Proclus Diadochus (ca. 411–485) writes (in *A Commentary on the First Book of Euclid's Elements*): "Eudoxus . . . multiplied the number of theorems which Plato originated concerning the 'section,' " he may be referring to Plato's (and Eudoxus') association with the Golden Section. This interpretation, however, has been a matter of serious controversy since the second half of the nineteenth century, with many researchers concluding that the word "section" probably has nothing to do with the Golden Section but rather is referring to the section of solids or to the general sectioning of lines. Nevertheless, there is little doubt that much of the groundwork leading to the definition and understanding of the Golden Ratio was carried out during the years just prior to the opening of Plato's Academy in 386 B.C. and throughout the period of the Academy's operation. The key figure and driving force behind the geometrical theorems concerning the Golden Ratio was probably Theaetetus (ca. 417 B.C.–ca. 369 B.C.), who according to the Byzantine collection *Suidas* "was the first to construct the five so-called solids." The fourth-century mathematician Pappus tells us that Theaetetus was also the one to have "distinguished the powers which are commensurable in length from those which are incommensurable." Theaetetus was not attached to the Academy directly, but he surely had some informal connections with it.

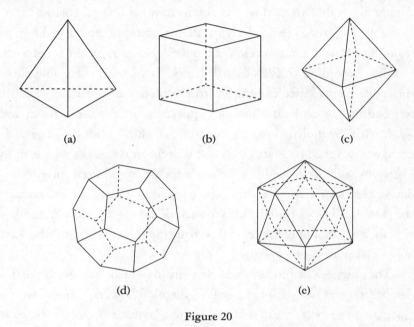

(a)   (b)   (c)

(d)   (e)

Figure 20

In *Timaeus*, Plato takes on the immense task of discussing the origin and workings of the cosmos. In particular, he attempts to explain the structure of matter using the five regular solids (or polyhedra), which had been investigated already to some extent by the Pythagoreans and very thoroughly by Theaetetus. The five Platonic solids (Figure 20) are distinguished by the following properties: They are the only existing solids in which all the faces (of a given solid) are identical and equilateral, and each of the solids can be circumscribed by a sphere (with all its vertices lying on the sphere). The Platonic solids are the tetrahedron (with four triangular faces; Figure 20a), the cube (with six square faces; Figure 20b), the octahedron (with eight triangular faces; Figure 20c);

the dodecahedron (with twelve pentagonal faces; Figure 20d), and the icosahedron (with twenty triangular faces; Figure 20e).

Plato combined the ideas of Empedocles (ca. 490–430 B.C.), that the four basic elements of matter are earth, water, air, and fire, with the "atomic" theory of matter (the existence of indivisible particles) of Democritus of Abdera (ca. 460 B.C.–ca. 370 B.C.). His "unified" theory suggested that each of the four elements corresponds to a different kind of fundamental particle and is represented by one of the Platonic solids. We should realize that while the details have, of course, changed considerably, the basic idea underlying Plato's theory is not that different from John Dalton's formulation of modern chemistry in the nineteenth century. According to Plato, Earth is associated with the stable cube, the "penetrating" quality of fire with the pointy and relatively simple tetrahedron, air with the "mobile" appearance of the octahedron, and water with the multifaceted icosahedron. The fifth solid, the dodecahedron, was assigned by Plato (in *Timaeus*) to the universe as a whole, or in his words, the dodecahedron is that "which the god used for embroidering the constellations on the whole heaven." This is the reason why the painter Salvador Dali decided to include a huge dodecahedron floating above the supper table in his painting "Sacrament of the Last Supper" (Figure 5 on page 9).

The absence of a fundamental element to be associated with the dodecahedron was not accepted by all of Plato's followers, some of whom postulated the existence of a fifth element. Aristotle, for example, took the aether, the material of heavenly bodies which he assumed to permeate the entire universe, to be the cosmic fifth essence ("quintessence"). He posited that by pervading all matter, this fifth essence ensured that motion and change could occur, in accordance with the laws of nature. The idea of a substance that pervades all space as a necessary medium for the propagation of light continued to hold until 1887, when a famous experiment by American physicist Albert Abraham Michelson and chemist Edward Williams Morley showed that this medium does not exist (nor is it required by the modern theory of light). Basically, the experiment measured the speed of two beams of light launched in different directions. The expectation was that because of Earth's motion through the aether, the speeds of the two beams would be different, but the experiment proved categorically that they were not. The result of the Michelson-Morley experiment set Einstein on the road to the theory of relativity.

In a surprising turn of events, in 1998 two groups of astronomers discovered that not only is our universe expanding (a fact already discovered by astronomer Edwin Hubble in the 1920s), but that the expansion is *accelerating*. This discovery came as a total shock, since astronomers naturally assumed that, due to gravity, the expansion should be slowing down. In the same way that a ball thrown upward on Earth continuously slows down because of gravity's pull (and eventually reverses its motion), the gravitational force exerted by all the matter in the universe should cause the cosmic expansion to decelerate. The discovery that the expansion is speeding up rather than slowing down suggests the existence of some form of "dark energy" that manifests itself as a repulsive force, which in our present-day universe overcomes the attractive force of gravity. Physicists are still struggling to understand the source and nature of this "dark energy." One suggestion is that this energy is associated with some quantum mechanical field that permeates the cosmos, a bit like the familiar electromagnetic field. Borrowing from Aristotle's invisible medium, this field has been dubbed "quintessence." Incidentally, in Luc Besson's 1997 science fiction movie *The Fifth Element,* the "fifth element" of the title was taken to be the life force itself—that which animates the inanimate.

Plato's theory was much more than a symbolic association. He noted that the faces of the first four solids could be constructed out of two types of right-angled triangles, the isosceles 45°-90°-45° triangle and the 30°-90°-60° triangle. Plato went on to explain how basic "chemical reactions" could be described using these properties. For example, in Plato's "chemistry," when water is heated by fire, it produces two particles of vapor (air) and one particle of fire. In a chemical reaction formulation, this may be written as

$$[water] \rightarrow 2[air] + [fire]$$

or, in balancing the number of faces involved (in the Platonic solids that represent these elements, respectively): $20 = 2 \times 8 + 4$. While this description clearly does not conform with our modern understanding of the structure of matter, the central idea—that the most fundamental particles in our universe and their interactions can be described by a mathematical theory that possesses certain symmetries—is one of the cornerstones of today's research in particle physics.

To Plato, the complex phenomena that we observe in the universe are not what really matters; the truly fundamental things are the underlying symmetries, and those are never changing. This view is very much in line with modern thinking about the laws of nature. For example, these laws do not change from place to place in the universe. For this reason, we can use the same laws that we determine from laboratory experiments whether we study a hydrogen atom here on Earth or in a galaxy that is billions of light-years away. This symmetry of the laws of nature manifests itself in the fact that the quantity which we call *linear momentum* (equaling the product of the mass of an object and the speed, and having the direction of the motion) is conserved, namely, has the same value whether we measure it today or a year from now. Similarly, because the laws of nature do not change with the passing of time, the quantity we call energy is conserved. We cannot get energy out of nothing. Modern theories, which are based on symmetries and conservation laws, are thus truly Platonic.

The original fascination of the Pythagoreans with polyhedra may have originated from observations of pyrite crystals in southern Italy, where the Pythagorean school was located. Pyrite, commonly known as fool's gold, often has crystals with a dodecahedral shape. However, the Platonic solids, their beauty, and their mathematical properties continued to captivate the imagination of people for centuries after Plato, and they turn up in the most unexpected places. For example, in Cyrano de Bergerac's (1619–1655) science-fiction novel *A Voyage to the Moon: with Some Account of the Solar World,* the author uses a flying machine in the form of an icosahedron to escape from prison in a tower and to land on a sunspot.

The Golden Ratio, ϕ, plays a crucial role in the dimensions and symmetry properties of some Platonic solids. In particular, a dodecahedron with an edge length (the segment along which two faces join) of one unit has a total surface area of $15\phi/\sqrt{3 - \phi}$ and a volume of

$5\phi^3/(6-2\phi)$. Similarly, an icosahedron with a unit length edge has a volume of $5\phi^5/6$.

The symmetry of the Platonic solids leads to other interesting properties. For example, the cube and the octahedron have the same number of edges (twelve), but their number of faces and vertices are interchanged (the cube has six faces and eight vertices and the octahedron eight faces and six vertices). The same is true for the dodecahedron and icosahedron; both have thirty edges, and the dodecahedron has twelve faces and twenty vertices, while it is the other way around for the icosahedron. These similarities in the symmetries of the Platonic solids allow for interesting mappings of one solid into its dual or reciprocal solid. If we connect the centers of all the faces of a cube, we obtain an octahedron (Figure 21), while if we connect the centers of the faces of an octahedron, we obtain a cube. The same procedure can be applied to map an icosahedron into a dodecahedron and vice versa, and the ratio of the edge lengths of

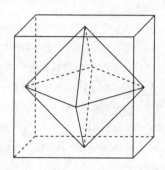

Figure 21

the two solids (one embedded in the other) that are obtained can again be expressed in terms of the Golden Ratio, as $\phi^2/\sqrt{5}$. The tetrahedron is self-reciprocating—joining the four centers of the tetrahedron's faces makes another tetrahedron.

While not all the properties of the Platonic solids were known in antiquity, neither Plato nor his followers failed to see their sheer beauty. To some extent, even the initial difficulties in constructing these figures (until methods using the Golden Ratio were found) could be taken as their attributes. After all, the last sentence in *Hippias Major* reads: "All that is beautiful is difficult." In "On the Failure of the Oracles," the Greek historian Plutarch (ca. 46–ca. 120) writes: "A pyramid [a tetrahedron], an octahedron, an icosahedron, and a dodecahedron, the primary figures which Plato predicates, are all beautiful because of the symmetries and equalities in their relations, and nothing superior or even like to these has been left for Nature to compose and fit together."

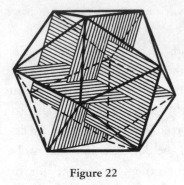

Figure 22

As noted above, the icosahedron and the dodecahedron are intimately related to the Golden Ratio, in more ways than one. For example, the twelve vertices of any icosahedron can be divided into three groups of four, with the vertices of each group lying at the corners of a *Golden Rectangle* (a rectangle in which the ratio of length to width is the Golden Ratio). The rectangles are perpendicular to each other, and their one common point is the center of the icosahedron (Figure 22). Similarly, the centers of the twelve pentagonal faces of the dodecahedron can be divided into three groups of four, and each of those groups also forms a Golden Rectangle.

The close associations between some plane figures, such as the pentagon and the pentagram, and some solids, such as the Platonic solids, and the Golden Ratio lead to the inescapable conclusion that the Greek interest in the Golden Ratio probably started with attempts to construct such plane figures and solids. Most of this mathematical effort occurred around the beginning of the fourth century B.C. There exist, however, numerous claims that the Golden Ratio is embodied in the architectural design of the Parthenon, which was built and decorated between 447 and 432 B.C., under the rule of Pericles. Can these claims be verified?

THE VIRGIN'S PLACE

The Parthenon ("the virgin's place" in Greek) was built on the Acropolis in Athens as a temple sacred to the cult of Athena Parthenos (Athena the Virgin). The architects were Ictinus and Callicrates, and Phidias and his assistants and students were charged with supervising the sculptures. Sculptured groups ornamented the pediments terminating the roof at the eastern and western ends. One group depicted the birth of Athena and the other the contest between Athena and Poseidon.

Somewhat deceptive in its simplicity, the Parthenon remains one of the finest architectural expressions of the ideal of clarity and unity. On September 26, 1687, Venetian artillery hit the Parthenon directly, during an attack on the Ottoman Turks who held Athens at the time and who used the Parthenon as a powder magazine. While the damage was extensive, the basic structure remained intact. In describing this event, General Königsmark, who accompanied the field commander, wrote: "How it dismayed His Excellency to destroy the beautiful temple which had existed three thousand years!" Numerous attempts have been made, especially since the end of the Turkish control (in 1830), to discover some mathematical or geometrical basis supposedly employed to achieve the Parthenon design's high perfection. Most books on the Golden Ratio state that the dimensions of the Parthenon, while its triangular pediment was still intact, fit precisely into a Golden Rectangle. This statement is usually accompanied by a figure similar to that in Figure 23. The Golden Ratio is assumed to feature in other dimensions of the Parthenon as well. For example, in one of the most extensive works on the Golden Ratio, Adolph Zeising's *Der Goldne Schnitt*

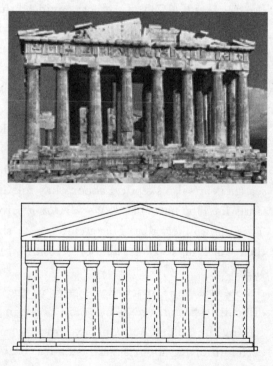

(The golden section; published in 1884), Zeising claims that the height of the façade from the top of its tympanum to the bottom of the pedestal below the columns is also divided in a Golden Ratio by the top of the columns. This statement was repeated in many books, such as Matila Ghyka's influential *Le Nombre d'or* (The golden number; appeared in 1931). Other authors, such as Miloutine Borissavlievitch in

Figure 23

The Golden Number and the Scientific Aesthetics of Architecture (1958), while not denying the presence of φ in the Parthenon's design, suggest that the temple owes its harmony and beauty more to the regular rhythm introduced by the repetition of the same column (a concept termed the "law of the Same").

The appearance of the Golden Ratio in the Parthenon was seriously questioned by University of Maine mathematician George Markowsky in his 1992 *College Mathematics Journal* article "Misconceptions about the Golden Ratio." Markowsky first points out that invariably, parts of the Parthenon (e.g., the edges of the pedestal; Figure 23) actually fall outside the sketched Golden Rectangle, a fact totally ignored by all the Golden Ratio enthusiasts. More important, the dimensions of the Parthenon vary from source to source, probably because different reference points are used in the measurements. This is another example of the number-juggling opportunity afforded by claims based on measured dimensions alone. Using the numbers quoted by Marvin Trachtenberg and Isabelle Hyman in their book *Architecture: From Prehistory to Post-Modernism* (1985), I am not convinced that the Parthenon has anything to do with the Golden Ratio. These authors give the height as 45 feet 1 inch and the width as 101 feet 3.75 inches. These dimensions give a ratio of width/height of approximately 2.25, far from the Golden Ratio of 1.618 . . . Markowsky points out that even if we were to take the height of the apex above the pedestal upon which the series of columns stands (given as 59 feet by Stuart Rossiter in his 1977 book *Greece*), we still would obtain a width/height ratio of about 1.72, which is closer to but still significantly different from the value of φ. Other researchers are also skeptical about phi's role in the Parthenon's design. Christine Flon notes in *The World Atlas of Architecture* (1984) that while "it is not unlikely that some architects . . . should have wished to base their works on a strict system of ratios . . . it would be wrong to generalize."

So, was the Golden Ratio used in the Parthenon's design? It is difficult to say for sure. While most of the mathematical theorems concerning the Golden Ratio (or "extreme and mean ratio") appear to have been formulated after the Parthenon had been constructed, considerable knowledge existed among the Pythagoreans prior to that. Thus, the

Parthenon's architects might have decided to base its design on some prevailing notion for a canon for aesthetics. However, this is far less certain than many books would like us to believe and is not particularly well supported by the actual dimensions of the Parthenon.

Whether or not the Golden Ratio features in the Parthenon, what is clear is that whichever mathematical "programs" concerning the Golden Ratio were instituted by the Greeks in the fourth century B.C., that work culminated in the publication of Euclid's *Elements,* in around 300 B.C. Indeed, from a perspective of logic and rigor, the *Elements* was long thought to be an apotheosis of certainty in human knowledge.

EXTREME AND MEAN RATIO

In 336 B.C., twenty-year-old Alexander (the Great) of Macedonia succeeded to the throne and, by a sequence of brilliant victories, conquered most of Asia Minor, Syria, Egypt, and Babylon and became ruler of the Persian Empire. A few years before his death at the young age of thirty-three, he founded what became the greatest monument to his name—the city of Alexandria near the mouth of the Nile.

Alexandria was located at the crossroads of three great civilizations: Egyptian, Greek, and Jewish. Consequently, it became an extraordinary intellectual center that lasted for centuries and the birthplace of such remarkable achievements as the Septuagint (meaning "translation of the 70")—the Greek translation of the Old Testament, traditionally attributed to seventy-two translators. The translation was begun in the third century B.C., and the work progressed in several stages over about a century.

After Alexander's death, Ptolemy I gained control over Egypt and the African dominions by 306 B.C., and among his first actions was the establishment of the equivalent of a university (known then as the Museum) in Alexandria. This institution included a library, which, following an immense gathering effort, was reputed to hold at one time 700,000 books (some confiscated from unlucky tourists). The first staff of teachers at the Alexandria school included Euclid, the author of the best-known book in the history of mathematics—the *Elements (Stoichia)*. In spite of Euclid being a "best-selling" author (only the Bible sold more books than *Elements* until the twentieth century), his life is so veiled in obscurity that even his birthplace in unknown. Given the contents of the *Elements,* it is very likely that Euclid studied mathematics in Athens with some of Plato's students. Indeed, Proclus writes about Euclid: "This man lived in the time of the first Ptolemy . . . he is then younger than the pupils of Plato, but older than Eratosthenes and Archimedes."

The *Elements,* a thirteen-volume work on geometry and number theory, is so colossal in its scope that we sometimes tend to forget that Euclid was the author of almost a dozen other books, covering topics from music through mechanics to optics. Only four of these other treatises survived to the present day: *Division of Figures, Optics, Phaenomena,* and *Data. Optics* contains some of the earliest studies of perspective.

Few will disagree that the *Elements* is the greatest and most influential mathematical textbook ever written. A story has it that when Abraham Lincoln wanted to understand the true meaning of the word "proof" in the legal profession, he started to study the *Elements* in his cabin in Kentucky. The famous British logician and philosopher Bertrand Russell describes in his *Autobiography* his first encounter with Euclid's *Elements* (at age eleven!) as "one of the great events of my life, as dazzling as first love."

The picture of the author that emerges from the pages of the *Elements* is that of a conscientious man, respectful of tradition, and very modest. Nowhere does Euclid attempt to take credit for work that was not originally his. In fact, he claims no originality whatsoever, in spite of the fact that it is very obvious that he contributed many new proofs, totally rearranged the contents contributed by others to entire volumes, and designed the whole work. Euclid's scrupulous fairness and modesty gained him the admiration of Pappus of Alexandria, who around A.D. 340 composed an eight-volume work entitled *Collection (Synagoge),* which provides an invaluable record of many aspects of Greek mathematics.

In the *Elements,* Euclid attempted to encompass most of the mathematical knowledge of his time. Books I to VI deal with the plane geometry we learn in school and which has become synonymous with Euclid's name (Euclidean geometry). Of these books, I, II, IV, and VI discuss lines and plane figures, while Book III presents theorems related to the circle, and Book V gives an extensive account of the work on proportion originated by Eudoxus of Cnidus (ca. 408–355 B.C.). Books VII to X deal with number theory and the foundations of arithmetic. In particular, irrational numbers are elaborated on in Book X, the contents of which are mostly the work of Theaetetus. Book XI provides the basis for solid geometry; Book XII (mostly describing the work of Eudoxus) proves the theorem for the area of the circle, and Book XIII (due mostly to Theaetetus) demonstrates the constructions of the five Platonic solids.

Still in ancient times, Hero (in the first century A.D.), Pappus (in the fourth century), and Proclus (in the fifth century), all of Alexandria, and Simplicius of Athens (in the sixth century) all wrote commentaries on the *Elements.* A new revision of the work, by Theon of Alexandria, appeared in the fourth century A.D. and served as the basis for all translations until the nineteenth century, when a manuscript containing a somewhat different text was discovered in the Vatican. In the Middle Ages, the *Elements* was translated into Arabic three times. The first of these translations was carried out by al-Hajjāj ibn Yūsuf ibn Maṭar, at the request of Caliph Hārūn ar-Rashīd (ruled 786–809), who is familiar to us through the stories in *The Arabian Nights.* The *Elements* was first made known in Western Europe through Latin translations of the Arabic versions. English Benedictine monk Adelard of Bath (ca. 1070–1145), who according to some stories was traveling in Spain disguised as a Muslim student, got hold of an Arabic text and completed the translation into Latin around 1120. This translation became the basis of all editions in Europe until the sixteenth century. Translations into many modern languages followed.

While Euclid himself may not have been the greatest mathematician who ever lived, he was certainly the greatest teacher of mathematics. The textbook he wrote remained in use practically unaltered for more than two thousand years, until the middle of the nineteenth century. Even the fictional detective Sherlock Holmes, in Arthur Conan Doyle's *A Study in Scarlet,* claimed that his conclusions, achieved by deduction, were "as infallible as so many propositions of Euclid."

The Golden Ratio appears in the *Elements* in several places. The first definition of the Golden Ratio ("extreme and mean ratio"), in relation to areas, is given somewhat obliquely in Book II. A second, clearer definition, in relation to proportion, appears in Book VI. Euclid then uses the Golden Ratio, especially in the construction of the pentagon (in Book IV) and in the construction of the icosahedron and dodecahedron (in Book XIII).

A ————————————— x ——————————— C ———— 1 ———— B

Figure 24

Let me use some very simple geometry to examine Euclid's definition and explain why the Golden Ratio is so important for the construction of the pentagon. In Figure 24, the line AB is divided by point C. Euclid's definition in Book VI of extreme and mean ratio is such that: (larger segment)/(shorter segment) is equal to (whole line/larger segment). In other words, in Figure 24:

$$AC/CB = AB/AC.$$

How is this line division related to the pentagon? In any regular planar figure (those with equal sides and interior angles; known as regular *polygons*), the sum of all the interior angles is given by 180 $(n - 2)$, where n is the number of sides. For example, in a triangle $n = 3$, and the sum of all the angles is equal to 180 degrees. In a pentagon $n = 5$, and the sum of all the angles is equal to 540 degrees. Every angle of the pentagon is therefore equal to 540/5 = 108 degrees. Imagine now that we draw two adjacent diagonals in the pentagon, as in Figure 25a, thus forming three isosceles (with two equal sides) triangles. Since the two angles near the base of an isosceles triangle are equal, the base angles in the two triangles on the sides are 36 degrees each [half of (180° − 108°)]. We therefore obtain for the angles of the middle triangle the values 36-72-72 (as marked in Figure 25a). If we bisect one of the two 72-degree base angles (as in Figure 25b), we obtain a smaller triangle DBC with the same angles (36-72-72) as the large one ADB. Using very elementary geometry, we can show that according to Euclid's definition, point C divides the line AB precisely in a Golden Ratio. Furthermore, the ratio of AD to DB is also equal to the Golden Ratio. (A short proof is given in Appendix 4.) In other words, in a regular pentagon, the ratio of the diagonal to the side is equal to φ. This fact illustrates that the ability to construct a line divided in a Golden Ratio provides at the same time a simple means of constructing the regular pentagon. The construction of the pentagon was the main reason for the Greek interest in the Golden Ratio. The triangle in the middle of Figure 25a, with a ratio of side to base of φ, is known as a *Golden Triangle;* the two triangles on the sides, with a ratio of side to base of 1/φ, are sometimes called *Golden Gnomons.* Figure 25b demonstrates a unique property of Golden Triangles and Golden Gnomons—they can be dissected into smaller triangles that are also Golden Triangles and Golden Gnomons.

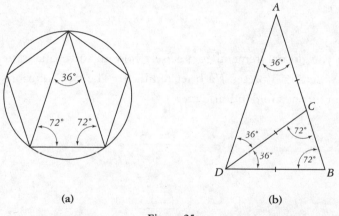

(a)   (b)

Figure 25

The association of the Golden Ratio with the pentagon, fivefold symmetry, and the Platonic solids is interesting in itself and, indeed, was more than sufficient to ignite the curiosity of the ancient Greeks. The Pythagorean fascination with the pentagon and the pentagram, coupled with Plato's interest in the regular solids and his belief that the latter represented fundamental cosmic entities, prompted generations of mathematicians to labor on the formulation of numerous theorems concerning φ. Yet the Golden Ratio would not have reached the level of almost reverential status that it eventually achieved were it not for some

truly unique algebraic properties. In order to understand these properties, we need first to find the precise value of φ.

    Examine again Figure 24, and let us take the length of the shorter
segment, *CB*, to be 1 unit and the length of the longer one, *AC*, to be *x*
units. If the ratio of *x* to 1 is the same as that of *x* + 1 (length of the line
AB) to *x*, then the line has been cut in extreme and mean ratio. We can
easily solve for the value, *x*, of the Golden Ratio. From the definition of
extreme and mean ratio

$$\frac{x}{1} = \frac{x+1}{x}$$

Multiplying both sides by *x*, we get $x^2 = x + 1$, or the simple quadratic
equation

$$x^2 - x - 1 = 0.$$

In case you do not remember precisely how to solve quadratic equations, Appendix 5 presents a brief reminder. The two solutions of the
equation for the Golden Ratio are:

$$x_1 = \frac{1 + \sqrt{5}}{2}$$

$$x_2 = \frac{1 - \sqrt{5}}{2} \; .$$

The positive solution $(1 + \sqrt{5})/2 = 1.6180339887\ldots$ gives the value
of the Golden Ratio. We now see clearly that φ is irrational, being simply half the sum of 1 and the square root of 5. Even before we go any
further, we can get a feeling that this number has some interesting
properties by using a simple scientific pocket calculator. Enter the number 1.6180339887 and hit the $[x^2]$ button. Do you see something
surprising? Now enter the number again, and this time hit the
$[1/x]$ button. Intriguing, isn't it? While the square of the number

1.6180339887 ... gives 2.6180339887 ..., its reciprocal ("one over") gives 0.6180339887 ..., all having precisely the same digits after the decimal point! The Golden Ratio has the unique properties that we produce its square by simply adding the number 1 and its reciprocal by subtracting the number 1. Incidentally, the negative solution of the equation $x_2 = (1 - \sqrt{5})/2$ is equal precisely to the negative of $1/\phi$.

Paul S. Bruckman of Concord, California, published in 1977 in the journal *The Fibonacci Quarterly* an amusing poem called "Constantly Mean." Referring to the Golden Ratio as the "Golden Mean," the first verse from that poem reads:

> *The golden mean is quite absurd;*
> *It's not your ordinary surd.*
> *If you invert it (this is fun!),*
> *You'll get itself, reduced by one;*
> *But if increased by unity,*
> *This yields its square, take it from me.*

The fact that we now have an algebraic expression for the Golden Ratio allows us, in principle, to calculate it to a high precision. This is precisely what M. Berg did in 1966, when he used 20 minutes on an IBM 1401 mainframe computer to calculate ϕ to the 4,599th decimal place. (The result was published in the *Fibonacci Quarterly*.) The same can be achieved today on almost any personal computer in less than two seconds. In fact, the Golden Ratio was computed to 10 million decimal places in December 1996, and it took about thirty minutes. For the true number enthusiasts, here is ϕ to the 2,000th decimal place:

									Decimal place
1.61803	39887	49894	84820	45868	34365	63811	77203	09179 80576	50
28621	35448	62270	52604	62818	90244	97072	07204	18939 11374	100
84754	08807	53868	91752	12663	38622	23536	93179	31800 60766	
72635	44333	89086	59593	95829	05638	32266	13199	28290 26788	200
06752	08766	89250	17116	96207	03222	10432	16269	54862 62963	
13614	43814	97587	01220	34080	58879	54454	74924	61856 95364	300
86444	92410	44320	77134	49470	49565	84678	85098	74339 44221	
25448	77066	47809	15884	60749	98871	24007	65217	05751 79788	400

```
34166  25624  94075  89069  70400  02812  10427  62177  11177  78053
15317  14101  17046  66599  14669  79873  17613  56006  70874  80710     500

13179  52368  94275  21948  43530  56783  00228  78569  97829  77834
78458  78228  91109  76250  03026  96156  17002  50464  33824  37764
86102  83831  26833  03724  29267  52631  16533  92473  16711  12115
88186  38513  31620  38400  52221  65791  28667  52946  54906  81131
71599  34323  59734  94985  09040  94762  13222  98101  72610  70596
11645  62990  98162  90555  20852  47903  52406  02017  27997  47175
34277  75927  78625  61943  20827  50513  12181  56285  51222  48093
94712  34145  17022  37358  05772  78616  00868  83829  52304  59264
78780  17889  92199  02707  76903  89532  19681  98615  14378  03149
97411  06926  08867  42962  26757  56052  31727  77520  35361  39362    1000

10767  38937  64556  06060  59216  58946  67595  51900  40055  59089
50229  53094  23124  82355  21221  24154  44006  47034  05657  34797
66397  23949  49946  58457  88730  39623  09037  50339  93856  21024
23690  25138  68041  45779  95698  12244  57471  78034  17312  64532
20416  39723  21340  44449  48730  23154  17676  89375  21030  68737
88034  41700  93954  40962  79558  98678  72320  95124  26893  55730
97045  09595  68440  17555  19881  92180  20640  52905  51893  49475
92600  73485  22821  01088  19464  45442  22318  89131  92946  89622
00230  14437  70269  92300  78030  85261  18075  45192  88770  50210
96842  49362  71359  25187  60777  88466  58361  50238  91349  33331

22310  53392  32136  24319  26372  89106  70503  39928  22652  63556
20902  97986  42472  75977  25655  08615  48754  35748  26471  81414
51270  00602  38901  62077  73224  49943  53088  99909  50168  03281
12194  32048  19643  87675  86331  47985  71911  39781  53978  07476
15077  22117  50826  94586  39320  45652  09896  98555  67814  10696
83728  84058  74610  33781  05444  39094  36835  83581  38113  11689
93855  57697  54841  49144  53415  09129  54070  05019  47754  86163
07542  26417  29394  68036  73198  05861  83391  83285  99130  39607
20144  55950  44977  92120  76124  78564  59161  60837  05949  87860
06970  18940  98864  00764  43617  09334  17270  91914  33650  13715    2000
```

Intriguing as they are, you may think that the properties of φ I have
described so far hardly justify adjectives like "Golden" or "Divine," and
you would be right. But this has been just a first glimpse of the wonders
to come.

SURPRISES GALORE

Everyone is familiar with the feeling we experience when we suddenly
recognize the face of an old friend at a party where we were convinced

we hardly know anyone. You may also have a similar emotional response when you go to an art exhibition and, upon turning a corner, find yourself suddenly facing one of your favorite paintings. The entire notion of a "surprise party" is in fact based on the pleasure and gratification many of us feel when confronted with such unexpected appearances. Mathematics and the Golden Ratio in particular provide a rich treasury of such surprises.

Imagine that we are trying to determine the value of the following unusual expression that involves square roots that go on forever:

$$\sqrt{1 + \sqrt{1 + \sqrt{1 + \sqrt{1 + \ldots}}}}.$$

How would we even go about finding the answer? One rather cumbersome way could be to start by calculating $\sqrt{1 + \sqrt{1}}$ (which is $\sqrt{2} = 1.414 \ldots$), then to calculate $\sqrt{1 + \sqrt{1 + \sqrt{1}}}$, and so on, hoping that the subsequent values will converge rapidly to some number. But there may be a shorter, more elegant method of calculation. Suppose we denote the value we are seeking by x. We therefore have

$$x = \sqrt{1 + \sqrt{1 + \sqrt{1 + \sqrt{1 + \ldots}}}}.$$

Now let us square both sides of this equation. The square of x is x^2, and the square of the expression on the right-hand side simply removes the outermost square root (by the definition of a square root). We therefore obtain

$$x^2 = 1 + \sqrt{1 + \sqrt{1 + \sqrt{1 + \ldots}}}.$$

However, note that because the second expression on the right-hand side goes on to infinity, it is actually equal to our original x. We therefore obtain the quadratic equation $x^2 = 1 + x$. But this is precisely the equation that defines the Golden Ratio! We therefore found that our endless expression is actually equal to ϕ.

Let us now look at a very different type of never-ending expression, this time involving fractions:

$$1 + \cfrac{1}{1 + \cfrac{1}{1 + \cfrac{1}{1 + \cfrac{1}{1 + \ldots}}}} \, .$$

This is a special case of mathematical entities known as *continued fractions,* which are used quite frequently in number theory. How would we compute the value of this continued fraction? Again, we could in principle truncate the series of 1s at successively higher points, hoping to find the limit to which the continued fraction converges. Based on our previous experience, however, we could at least start by denoting the value by x. Thus,

$$x = 1 + \cfrac{1}{1 + \cfrac{1}{1 + \cfrac{1}{1 + \cfrac{1}{1 + \ldots}}}} \, .$$

Note, however, that because the continued fraction goes on forever, the *denominator* of the second term on the right-hand side is in fact identical to x itself. We therefore have the equation

$$x = 1 + \frac{1}{x} \, .$$

Multiplying both sides by x, we get $x^2 = x + 1$, which is again the equation defining the Golden Ratio! We find that this remarkable continued fraction is also equal to ϕ. Paul S. Bruckman's poem "Constantly Mean" refers to this property as well:

> *Expressed as a continued fraction,*
> *It's one, one, one, . . . , until distraction;*
> *In short, the simplest of such kind*
> *(Doesn't this really blow your mind?)*

Because the continued fraction corresponding to the Golden Ratio is composed of ones only, it converges very slowly. The Golden Ratio is, in this sense, more "difficult" to express as a fraction than any other irrational number—it is the "most irrational" among irrationals.

From never-ending expressions let us now turn our attention to the Golden Rectangle in Figure 26. The lengths of the sides of the rectangle are in a Golden Ratio to each other. Suppose we cut off a square from this rectangle (as marked in the figure). We will be left with a smaller rectangle that is also a Golden Rectangle. The dimensions of the "daughter" rectangle are smaller than those of the "parent" rectangle by precisely a factor φ. We can now cut a square from the daughter Golden Rectangle and we will be left again with a Golden Rectangle, the dimensions of which are smaller by another factor of φ. Continuing this process ad infinitum, we will produce smaller and smaller Golden Rectangles (each time with dimensions "deflated" by a factor φ). If we were to examine the ever-decreasing (in size) rectangles with a magnifying glass of increasing power, they would all look identical. The Golden Rectangle is the *only* rectangle with the property that cutting a square from it produces a similar rectangle. Draw two diagonals of any mother-daughter pair of rectangles in the series, as in Figure 26, and they will all intersect at the same point. The series of continuously diminishing rectangles converges to that never-reachable point. Because of the "divine" properties attributed to the Golden Ratio, mathematician Clifford A. Pickover suggested that we should refer to that point as "the Eye of God."

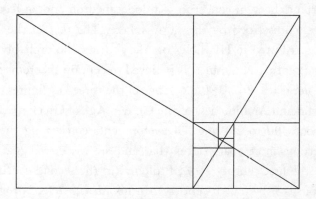

Figure 26

If you did not find it mind-boggling that all of these diverse mathematical circumstances lead to φ, take a simple pocket calculator and I will show you an amazing magic trick. Choose any two numbers (with any number of digits) and write them one after the other. Now, using the calculator (or in your head), form a third number, by simply adding together the first two (and write it down); form a fourth number by adding the second number to the third; a fifth number by adding the third to the fourth; a sixth number by adding the fourth to the fifth and so on, until you have a series of twenty numbers. For example, if your first two numbers were 2 and 5, you would have obtained the series 2, 5, 7, 12, 19, 31, 50, 81, 131. . . . Now use the calculator to divide your twentieth number by your nineteenth number. Does the result look familiar? It is, of course, phi. I shall return to this trick and its explanation in Chapter 5.

TOWARD THE DARK AGES

In his definition in the *Elements,* Euclid was interested primarily in the geometrical interpretation of the Golden Ratio and in its use in the construction of the pentagon and some Platonic solids. Following in his footsteps, Greek mathematicians in the next centuries produced several new geometrical results involving the Golden Ratio. For example, the "Supplement" to the *Elements* (often referred to as Book XIV) contains an important theorem concerning a dodecahedron and an icosahedron that are circumscribed by the same sphere. The text of the "Supplement" is attributed to Hypsicles of Alexandria, who probably lived in the second century B.C., but it is believed to contain theorems by Apollonius of Perga (ca. 262–190 B.C.), one of the three key figures (together with Euclid and Archimedes) of the Golden Age of Greek mathematics (from about 300 to 200 B.C.). Developments concerning the Golden Ratio become more sparse after that and are associated mainly with Hero (in the first century A.D.), Ptolemy (in the second century A.D.), and Pappus (in the fourth century). In his *Metrica,* Hero provided approximations (often without offering a clue of how they were obtained) for the areas of the pentagon and the decagon (the ten-sided polygon) and for the volumes of dodecahedrons and icosahedrons.

Ptolemy (Claudius Ptolemaus) lived around A.D. 100 to 179, but virtually nothing is known about his life, except that he did most of his work in Alexandria. Based on his own and previous astronomical observations, he developed his celebrated geocentric model of the universe, according to which the Sun and all the planets revolved around Earth. While fundamentally wrong, his model did manage to explain (at least initially) the observed motions of the planets, and it continued to govern astronomical thinking for some thirteen centuries.

Ptolemy synthesized his own astronomical work with that of other Greek astronomers (in particular Hipparchos of Nicaea) in an encyclopaedic, thirteen-volume book, *Hē Mathēmatikē Syntaxis* (The mathematical synthesis). The book later became known as *The Great Astronomer.* However, ninth-century Arab astronomers referred to the book invoking the Greek superlative "Megistē" ("the greatest") but prefixing it with the Arabic identifier of proper names, "al." The book thereby became known, to this day, as the *Almagest.* Ptolemy also did important work in geography and wrote an influential book entitled *Guide to Geography.*

In the *Almagest* and the *Guide to Geography,* Ptolemy constructed one of the earliest equivalents of a trigonometric table for many angles. In particular, he calculated lengths of chords connecting two points on a circle for various angles, including the angles 36, 72, and 108 degrees, which, as you recall, appear in the pentagon and are therefore closely associated with the Golden Ratio.

The last great Greek geometer who contributed theorems related to the Golden Ratio was Pappus of Alexandria. In his *Collection (Synagoge;* ca. A.D. 340), Pappus gives a new method for the construction of the dodecahedron and the icosahedron as well as comparisons of the volumes of the Platonic solids, all of which involve the Golden Ratio. Pappus' commentary on Euclid's theory of irrational numbers traces beautifully the historical development of irrationals and is extant in its Arabic translations. However, his heroic efforts to arrest the general decay of mathematics and of geometry in particular were unsuccessful, and after his death, with the overall withering of intellectual curiosity in the

West, interest in the Golden Ratio entered a long period of hibernation. The great Alexandrian library was destroyed by a series of attacks, first by the Romans and then by Christians and Muslims. Even Plato's Academy came to an end in A.D. 529, when the Byzantine emperor Justinian ordered the closing of all the Greek schools. During the depressing Dark Ages that followed, the French historian and bishop Gregory of Tours (538–594) lamented that "the study of letters is dead in our midst." In fact, the whole enterprise of science was essentially transferred in its entirety to India and the Arab world. A significant event of this period was the introduction of the so-called Hindu-Arabic numerals and of decimal notation. The most important Hindu mathematician of the sixth century was Āryabhaṭa (476–ca. 550). In his best-known book, entitled *Āryabhaṭíya,* we find the phrase "from place to place each is ten times the preceding," which indicates an application of a place-value system. An Indian plate from 595 already contains writing (of a date) in Hindu numerals using decimal place-value notation, implying that such numerals had been in use for some time. The first sign (albeit with no real influence) of Hindu numerals moving West can be found in the writings of the Nestorian bishop Severus Sebokht, who lived in Keneshra on the Euphrates River. He wrote in 662: "I will omit all discussion of the science of the Indians . . . and of their valuable methods of calculation which surpass description. I wish only to say that this computation is done by means of nine signs."

With the ascendancy of Islam, the Muslim world became an important center for mathematical study. Had it not been for the intellectual surge in Islam during the eighth century, most of the ancient mathematics would have been lost. In particular, Caliph al-Mamun (786–833) established in Baghdad the Beit al-hikma (House of wisdom), which operated in a similar fashion to the famous Alexandrian university or "Museum." Indeed, the Abbasid empire subsumed any Alexandrian learning that had survived. According to tradition, after having a dream in which Aristotle appeared, the caliph decided to have all the ancient Greek works translated.

Many of the important Islamic contributions were algebraic in nature and touched on the Golden Ratio only very peripherally. Nevertheless, at least three mathematicians should be mentioned: Al-Khwārizmī and Abu Kamil Shuja in the ninth century and Abu'l-Wafa in the tenth century.

Mohammed ibn-Musa al-Khwārizmī composed, in Baghdad (at about 825), what is considered to be the most influential algebraic work of the period—*Kitāb al-jabr wa al-muqābalah* (The science of restoration and reduction). From this title ("al-jabr") comes the word "algebra" that we use today, since this was the first textbook used in Europe on that subject matter. Furthermore, the word "algorithm," used for any special method for solving a mathematical problem using a collection of exact procedural steps, comes from a distortion of al-Khwārizmī's name. *The Science of Restoration* was synonymous with the theory of equations for a few hundred years. The equation required to solve one of the problems presented by al-Khwārizmī bears a close resemblance to the equation defining the Golden Ratio. Al-Khwārizmī says: "I have divided ten into two parts; I have multiplied the one by ten and the other by itself, and the products were the same." Al-Khwārizmī calls the unknown shai ("the thing"). Consequently, the first line in the description of the equation obtained (for the above problem) translates to: "you multiply *thing* by ten; it is ten *things.*" The equation one obtains, $10x = (10 - x)^2$, is the one for the smaller segment of a line of length 10 divided in a Golden Ratio. The question of whether al-Khwārizmī actually had the Golden Ratio in mind when posing this problem is a matter of some dispute. Under the influence of al-Khwārizmī's work, the unknown was called "res" in the early algebraic works in Latin, translated to "cosa" ("the thing") in Italian. Accordingly, algebra itself became known as "l'arte della cosa" ("the art of the thing"). Occasionally it was referred to as the "ars magna" ("the great art"), to distinguish it from what was considered as the lesser art of arithmetic.

The second Arab mathematician who made contributions related to the history of the Golden Ratio is Abu Kamil Shuja, known as al-Hasib al-Misri, meaning "the Calculator from Egypt." He was born around 850, probably in Egypt, and died at about 930. He wrote many books, some of which, including the *Book on Algebra, Book of Rare Things in the Art of Calculation,* and *Book on Surveying and Geometry,* have survived. Abu Kamil may have been the first mathematician who instead of simply finding a solution to a problem was interested in finding *all* the possible solutions. In his *Book of Rare Things in the Art of Calculation* he even describes one problem for which he found 2,678 solutions. More important from the point of view of the history of the Golden Ratio, Abu Kamil's books served as the basis for some of the books of the Italian mathematician Leonardo of Pisa, known as Fibonacci, whom we shall encounter shortly. Abu Kamil's treatise *On the Pentagon and the Decagon* contains twenty problems and their solutions, in which he calculates the areas of the figures and the length of their sides and the radii of surrounding circles. In some of these calculations (but not all), he uses the Golden Ratio. A few of the algebraic problems appearing in *Algebra* may have also been inspired by the concept of the Golden Ratio.

The last of the Islamic mathematicians I would like to mention is Mohammed Abu'l-Wafa (940–998). Abu'l-Wafa was born in Buzjan (in present-day Iran) and lived during the rule of the Buyid Islamic dynasty in western Iran and Iraq. This dynasty reached its peak under the reign of Ádud ad-Dawlah, who was a great patron of mathematics, the sciences, and the arts. Abu'l-Wafa was one of the mathematicians who were invited to Ádud ad-Dawlah's court in Baghdad in 959. His first major book was *Book on What Is Needed from the Science of Arithmetic for Scribes and Businessmen,* and according to Abu'l-Wafa, it "comprises all that an experienced or novice, subordinate or chief in arithmetic needs to know." Interestingly, although Abu'l-Wafa himself was an expert in the use of Hindu numerals, all the text of his book is written with no numerals whatsoever—numbers are written only as words, and calculations are done only mentally. By the tenth century, the use of Indian numerals had not yet found application in the business circles. Abu'l-Wafa's interest in the Golden Ratio appears in his other book: *A Book on the Geometric Constructions Which Are Needed for an Artisan.* In this book, Abu'l-Wafa presents ingenious methods for the construction of the pentagon and the decagon and for inscribing regular polygons in circles and inside other polygons. A unique component of his work is a series of problems that he solves using a ruler (straightedge) and a compass, in which the angle between the two legs of the compass is kept fixed (known as "rusty compass" constructions). This particular genre was probably inspired by Pappus' *Collection* but may also represent Abu'l-

Wafa's response to a practical problem—the results with a fixed-angle compass are more accurate.

The work by these and other Islamic mathematicians produced important but only incremental progress in the mathematical history of the Golden Ratio. As is often the case in the sciences, such preparatory periods of slow advancement are necessary to give birth to the next breakthrough. The great playwright George Bernard Shaw once expressed his views on progress by the statement: "The reasonable man adapts himself to the world; the unreasonable one persists in trying to adapt the world to himself. Therefore all progress depends on the unreasonable man." In the case of the Golden Ratio, the quantum leap had to await the appearance of the most distinguished European mathematician of the Middle Ages—Leonardo of Pisa.

5

SON OF
GOOD NATURE

The nine Indian figures are: 9 8 7 6 5 4 3 2 1.
With these nine figures, and with the sign 0 . . . any number
may be written, as is demonstrated below.
—LEONARDO FIBONACCI (CA. 1170S–1240S)

With the above words, Leonardo of Pisa (in Latin Leonardus Pisanus), also known as Leonardo Fibonacci, began his first and best-known book, *Liber abaci* (Book of the abacus), published in 1202. At the time the book appeared, only a few privileged European intellectuals who cared to study the translations of the works of al-Khwārizmī and Abu Kamil knew the Hindu-Arabic numerals we use today. Fibonacci, who for a while joined his father, a customs and trading official, in Bugia (in present-day Algeria) and later traveled to other Mediterranean countries (including Greece, Egypt, and Syria), had the opportunity to study and compare different numerical systems and methods for arithmetical operations. Upon concluding that the Hindu-Arabic numerals, which included the place-value principle, were far superior to all other methods, he devoted the first seven chapters of his book to explanations of Hindu-Arabic notation and its use in practical applications.

Leonardo Fibonacci was born in the 1170s to a businessman and government official named Guglielmo. The nickname Fibonacci (from the Latin filius Bonacci, son of the Bonacci family, or "son of good nature") was most probably introduced by the historian of mathematics Guillaume Libri in a footnote in his 1838 book *Histoire des Sciences Mathematique en Italie* (History of the mathematical sciences in Italy), although some researchers attribute the first use of Fibonacci to Italian mathematicians at the end of the eighteenth century. In some manuscripts and documents, Leonardo either refers to himself or is referred to as Leonardo Bigollo (or Leonardi Bigolli Pisani), where "Bigollo" means something like "a traveler" or a "man of no importance" in the Tuscan and Venetian dialects respectively. Pisa of the twelfth century was a busy port through which merchandise passed both from inland and from overseas. Spices from the Far East circulated through Pisa on their way to northern Europe, crossing in the port the paths of wine, oil, and salt that were transported between different parts of Italy, Sicily, and Sardinia. The large Pisan leather industry imported goatskins from North Africa, and tanners could be seen processing hides on Pisa's riverbanks. The city, on the river Arno, was also proud of its excellent ironwork and shipyards. Pisa is best known today for its famous leaning tower, and the construction of this bell tower began during Fibonacci's youth. Clearly, all of this commercial frenzy required massive records of inventories and prices. Leonardo surely had the opportunity to watch various scribes as they were listing prices in Roman numerals and adding them up using an abacus. Arithmetic operations with Roman numerals are not fun. For example, to obtain the sum of 3,786 and 3,843, you would need to add MMMDCCLXXXVI to MMMDCC-CXLIII; if you think that is cumbersome, try multiplying those numbers. However, for as long as medieval merchants stuck to simple additions and subtractions, they could get by with Roman numerals. The fundamental element that the Roman numerals were lacking was, of course, the place-value system—the fact that a number written as 547 really means $(5 \times 10^2) + (4 \times 10^1) + (7 \times 10^0)$. The West Europeans overcame the lack of a place-value principle in their number system by the use of the abacus. The name "abacus" may have originated from *avaq,* the Hebrew word for dust, since the earliest of these calculation devices were simply boards dusted with sand on which numbers could be traced. The abacus during Fibonacci's time had counters sliding along wires. The different wires of the abacus played the role of place value. A typical abacus had four wires, with beads on the bottom wire representing units, those on the one above it tens, those on the third hundreds, and those on the top wire thousands. Thus, while the abacus provided a fairly efficient means for simple arithmetic operations (I was amazed to discover during a visit to Moscow in 1990 that the cafeteria in my hotel was still using an abacus), it clearly presented enormous disadvantages when handling more complex computations. It is impossible to imagine, for example, trying to manipulate the "billions and billions" of astronomy popularizer Carl Sagan using an abacus.

In Bugia (now called Bejaïa), in Algeria, Fibonacci became acquainted with the art of the nine Indian figures, probably with, in his words, the "excellent instruction" of an Arab teacher. Following a tour around the Mediterranean, which he used to expand his mathematical horizons, he decided to publish a book that would introduce the use of Hindu-Arabic numerals more widely into commercial life. In this book, Fibonacci meticulously explains the translation from Roman numerals to the new system and the arithmetic operations with the new numerals. He gives numerous examples that demonstrate the application of his "new math" to a variety of problems ranging from business practices and the filling and emptying of cisterns to the motions of ships. At the beginning of the book, Fibonacci adds the following apology: "If by chance I have omitted anything more or less proper or necessary, I beg forgiveness, since there is no one who is without fault and circumspect in all matters."

In many cases, Fibonacci gave more than one version of the problem, and he demonstrated an astonishing versatility in the choice of several methods of solution. In addition, his algebra was often rhetorical, explaining in words the desired solution rather than solving explicit equations, as we would do today. Here is a nice example of one of the problems that appear in *Liber abaci* (as translated in the charming book *Leonard of Pisa and the New Mathematics of the Middle Ages* by Joseph and Frances Gies):

A man whose end was approaching summoned his sons and said:
"Divide my money as I shall prescribe." To his eldest son, he said,
"You are to have 1 bezant [a gold coin first struck at Byzantium]

and a seventh of what is left." To his second son he said, "Take 2 bezants and a seventh of what remains." To the third son, "You are to take 3 bezants and a seventh of what is left." Thus he gave each son 1 bezant more than the previous son and a seventh of what remained, and to the last son all that was left. After following their father's instructions with care, the sons found that they had shared their inheritance equally. How many sons were there, and how large was the estate?

For the interested reader, I present both the algebraic (modern) solution and Fibonacci's rhetorical solution to this problem in Appendix 6.

The *Liber abaci* brought Fibonacci considerable recognition, and his fame reached even the ears of the Roman emperor Frederick II, known as "Stupor Mundi" ("Wonder of the World") for his patronage of mathematics and the sciences. Fibonacci was invited to appear before the emperor in Pisa in the early 1220s and was presented with a series of what were considered to be very difficult mathematical problems, by Master Johannes of Palermo, one of the court mathematicians. One of the problems read as follows: "Find such a *rational* number [a whole number or a fraction] that when 5 is either added to or subtracted from its square, the result [in either case] is also the square of a rational number." Fibonacci solved all the problems using ingenious methods. He later described two of them in a short book called *Flos* (Flower) and used the one above in the prologue of a book he dedicated to the emperor: *Liber quadratorum* (Book of squares). Today we have to be impressed by the fact that without relying on computers or calculators of any sort, simply through his virtuosic command of number theory, Fibonacci was able to find out that the solution to the problem above is $^{41}/_{12}$. Indeed, $(^{41}/_{12})^2 + 5 = (^{49}/_{12})^2$ and $(^{41}/_{12})^2 - 5 = (^{31}/_{12})^2$.

Fibonacci's role in the history of the Golden Ratio is truly fascinating. On one hand, in problems in which he consciously used the Golden Ratio, he is responsible for a significant but not spectacular progress. On the other, by simply formulating a problem that on the face of it has no relation whatsoever to the Golden Ratio, he expanded the scope of the Golden Ratio and its applications dramatically.

Fibonacci's direct contributions to the Golden Ratio literature appear in a short book on geometry, *Practica Geometriae* (Practice of geometry), which was published in 1223. He presented new methods for the calculation of the diagonal and the area of the pentagon, calculations of the sides of the pentagon and the decagon from the diameter of both inscribed and circumscribed circles, and computations of volumes of the dodecahedron and the icosahedron, all of which are intimately related to the Golden Ratio. In the solutions to these problems Fibonacci exhibits a deep understanding of Euclidean geometry. While his mathematical techniques rely to some extent on previous works, in particular on Abu Kamil's *On the Pentagon and the Decagon,* there is little doubt that Fibonacci brought the use of the Golden Ratio's properties in various geometrical applications to a higher level. However, Fibonacci's main claim to fame and his most exciting contribution to the Golden Ratio derive from an innocent-looking problem in *Liber abaci.*

ALL THE THOUGHTS OF A RABBIT ARE RABBITS

Many students of mathematics, the sciences, and the arts have heard of Fibonacci only because of the following problem from Chapter XII in the *Liber abaci.*

> A certain man put a pair of rabbits in a place surrounded on all sides by a wall. How many pairs of rabbits can be produced from that pair in a year if it is supposed that every month each pair begets a new pair which from the second month on becomes productive?

How can the numbers of the offspring of rabbits have significant mathematical consequences? Indeed, the solution to the problem itself is quite simple. We start with one pair. After the first month, the first pair gives birth to another pair, hence there are two. In Figure 27 I represent a mature pair with a large rabbit symbol and a young pair with a small symbol. After the second month, the mature pair gives birth to another young pair, while the baby pair matures. Hence, there are three pairs, as depicted in the figure. After the third month, each of the two mature pairs gives birth to another pair, and the baby pair matures, so there are five. After the fourth month, each of the three mature pairs gives birth to a pair, and the two baby pairs mature, giving us a total of eight pairs. After five months we have a baby pair from each of the five adult pairs, plus three maturing pairs for a total of thirteen. By now we understand how to proceed to obtain the numbers of mature pairs, of baby pairs, and of pairs in total in successive months. Suppose we examine just the number of adult pairs in any particular month. That number is composed of the number of adult pairs in the previous month, plus the number of baby pairs (which have matured) from that same previous month. However, the number of baby pairs from the previous month is actually equal to the number of adult pairs in the month before that. Therefore, in any given month (starting with the third), the number of adult pairs is simply equal to the sum of the numbers of adult pairs in the two preceding months. The number of adult pairs therefore follows the sequence: 1, 1, 2, 3, 5, 8, . . . You can easily see from the figure that the numbers of baby pairs follow precisely the same sequence, only displaced by one month. Namely, the numbers of baby pairs are 0, 1, 1, 2, 3, 5, 8, . . . Of course, the total number of pairs is simply the sum of these, and it gives the same sequence as for the adult pairs, with the first term omitted (1, 2, 3, 5, 8, . . .). The sequence 1, 1, 2, 3, 5, 8, 13, 21, 34, 55, 89, 144, 233, . . . , in which each term (starting with the third) is equal to the sum of the two preceding terms, was appropriately dubbed the Fibonacci sequence in the nineteenth century, by the French mathematician Edouard Lucas (1842–1891). Number sequences in which the relation between successive terms can be expressed by a mathematical expression are known as recursive. The Fibonacci sequence was the first such recursive sequence known in Europe. The general property that each term in the

Figure 27

sequence is equal to the sum of the two preceding ones is expressed mathematically as (a notation introduced in 1634 by the mathematician Albert Girard): $F_{n+2} = F_{n+1} + F_n$. Here F_n represents the nth number in the sequence (e.g., F_5 is the fifth term); F_{n+1} is the term following F_n (for $n = 5$, $n + 1 = 6$), and F_{n+2} follows F_{n+1}.

The reason that Fibonacci's name is so famous today is that the appearance of the Fibonacci sequence is far from being confined to the breeding of rabbits. Incidentally, the title of this chapter was inspired by Ralph Waldo Emerson's *The Natural History of Intellect,* which appeared in 1893. Emerson says: "All the thoughts of a turtle are turtles, and of a rabbit, rabbits." We shall encounter the Fibonacci sequence in an incredible variety of seemingly unrelated phenomena.

To start things off, let us examine a phenomenon that is just about as remote from the topic of rabbit progeny as we could possibly imagine—the optics of light rays. Suppose we have two glass plates made of slightly different types of glass (different light refraction properties, or "indices of refraction") mounted face to face (as in Figure 28a). If we shine light through the plates, the light rays can (in principle) reflect internally at four reflective surfaces before emerging (Figure 28a). More specifically, they can either pass through without reflecting at all, or they can undergo one internal reflection, two internal reflections, three internal reflections, and so on, potentially an infinite number of internal reflections before reemerging. All of these are paths allowed by the laws of optics. Now count the number of beams that emerge from this two-plate system. There is only one emerging beam in the case of no reflections at all (Figure 28b). There are two emerging beams when all the possibilities for the rays to undergo precisely one internal reflection are considered (Figure 28c), because there are two paths the ray can follow. There are three emerging beams for all the possibilities of two internal reflections (Figure 28d); five beams for three internal reflections (Figure 28e); eight paths if the ray is reflected four times (Figure 28f); thirteen paths for five reflections (Figure 28g); and so on. The numbers of emerging beams—1, 2, 3, 5, 8, 13 . . . —form a Fibonacci sequence.

Now consider the following entirely different problem. A child is trying to climb a staircase. The maximum number of steps he can climb at a

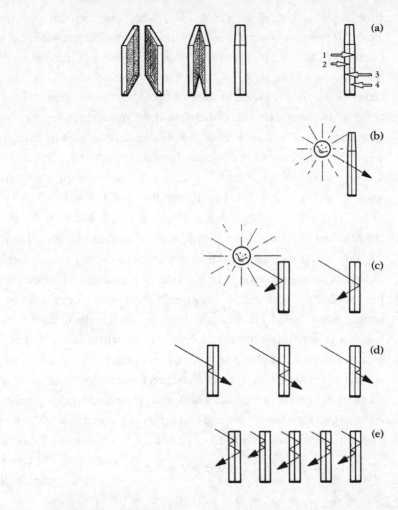

Figure 28

time is two; that is, he can climb either one step or two steps at a time. If there are n steps in total, in how many different ways, C_n, can he climb the staircase? If there is only one step $(n = 1)$, clearly there is only one way to climb it, $C_1 = 1$. If there are two steps, the child can either climb the two steps at once or take them one step at a time; thus, there are two ways, $C_2 = 2$. If there are three steps, there are three ways of climbing: $1 + 1 + 1$, $1 + 2$, or $2 + 1$; therefore $C_3 = 3$. If there are four steps, the number of ways to climb them increases to $C_4 = 5$: $1 + 1 + 1 + 1$, $1 + 2 + 1$, $1 + 1 + 2$, $2 + 1 + 1$, $2 + 2$. For five steps, there are eight ways, $C_5 = 8$: $1 + 1 + 1 + 1 + 1$, $1 + 1 + 1 + 2$, $1 + 1 + 2 + 1$, $1 + 2 + 1 + 1$, $2 + 1 + 1 + 1$, $2 + 2 + 1$, $2 + 1 + 2$, $1 + 2 + 2$. We find that the numbers of possibilities, $1, 2, 3, 5, 8, \ldots$, form a Fibonacci sequence.

Finally, let us examine the family tree of a drone, or male bee. Eggs of worker bees that are not fertilized develop into drones. Hence, a drone has no "father" and only a "mother." The queen's eggs, on the other hand, are fertilized by drones and develop into females (either workers or queens). A female bee has therefore both a "mother" and a "father." Consequently, one drone has one parent (its mother), two grandparents (its mother's parents), three great-grandparents (two parents of its grandmother and one of its grandfather), five great-great-grandparents (two for each great-grandmother and one for its great-grandfather), and so on. The numbers in the family tree, $1, 1, 2, 3, 5 \ldots$, form a Fibonacci sequence. The tree is presented graphically in Figure 29.

This all looks very intriguing—the same series of numbers applies to rabbits, to optics, to stair climbing, and to drone family trees—but how is the Fibonacci sequence related to the Golden Ratio?

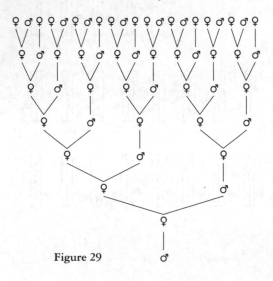

Figure 29

GOLDEN FIBONACCIS

Examine again the Fibonacci sequence; 1, 1, 2, 3, 5, 8, 13, 21, 34, 55, 89, 144, 233, 377, 610, 987, . . . , and this time let us look at the ratios of successive numbers (calculated here to the sixth decimal place):

$$1/1 = 1.000000$$
$$2/1 = 2.000000$$
$$3/2 = 1.500000$$
$$5/3 = 1.666666$$
$$8/5 = 1.600000$$
$$13/8 = 1.625000$$
$$21/13 = 1.615385$$
$$34/21 = 1.619048$$
$$55/34 = 1.617647$$
$$89/55 = 1.618182$$
$$144/89 = 1.617978$$
$$233/144 = 1.618056$$
$$377/233 = 1.618026$$
$$610/377 = 1.618037$$
$$987/610 = 1.618033$$

Do you recognize this last ratio? As we go farther and farther down the Fibonacci sequence, the ratio of two successive Fibonacci numbers oscillates about (being alternately greater or smaller) but comes closer and closer to the Golden Ratio. If we denote the nth Fibonacci number by F_n, and the next one by F_{n+1}, then we discovered that the ratio F_{n+1}/F_n approaches ϕ as n becomes larger. This property was discovered in 1611 (although possibly even earlier by an anonymous Italian) by the famous German astronomer Johannes Kepler, but more than a hundred years passed before the relation between Fibonacci numbers and the Golden Ratio was proven (and even then not fully) by the Scottish mathematician Robert Simson (1687–1768). Kepler, by the way, apparently hit upon the Fibonacci sequence on his own and not via reading the *Liber abaci*.

But why should the terms in a sequence derived from the breeding of rabbits approach a ratio defined through the division of a line? To understand this connection, we have to go back to the astonishing continued fraction we encountered in Chapter 4. Recall that we found that the Golden Ratio can be written as

$$\phi = 1 + \cfrac{1}{1 + \cfrac{1}{1 + \cfrac{1}{1 + \cfrac{1}{1 + \ldots}}}}.$$

In principle, we could calculate the value of ϕ by a series of successive approximations, in which we would interrupt the continued fraction farther and farther down. Suppose we attempted to do just that. We would find the series of values (reminder: 1 over a/b is equal to b/a):

$$1 = 1.00000$$

$$1 + \frac{1}{1} = \frac{2}{1} = 2.00000$$

$$1 + \frac{1}{1 + 1} = \frac{3}{2} = 1.50000$$

$$1 + \cfrac{1}{1 + \cfrac{1}{1 + 1}} = \frac{5}{3} = 1.66666$$

$$1 + \cfrac{1}{1 + \cfrac{1}{1 + \cfrac{1}{1 + 1}}} = \frac{8}{5} = 1.60000$$

$$1 + \cfrac{1}{1 + \cfrac{1}{1 + \cfrac{1}{1 + \cfrac{1}{1 + 1}}}} = \frac{13}{8} = 1.62500 .$$

In other words, the successive approximations we find for the Golden Ratio are precisely equal to the ratios of Fibonacci numbers. No wonder then that as we go to higher and higher terms in the sequence the ratio converges to the Golden Ratio. This property is described beautifully in the book *On Growth and Form* by the famous naturalist Sir D'Arcy Wentworth Thompson (1860–1948). He writes about the Fibonacci numbers: "Of these famous and fascinating numbers a mathematical friend writes to me: 'All the romance of continued fractions, linear recurrence relations, . . . lies in them, and they are a source of endless curiosity. How interesting it is to see them striving to attain the unattainable, the Golden Ratio, for instance; and this is only one of hundreds of such relations.' " The convergence to the Golden Ratio, by the way, explains the magic trick I described in Chapter 4. If you define a series of numbers by the property that each term (starting with the third) is equal to the sum of the two preceding ones, then irrespective of the two numbers you started with, as long as you go sufficiently far down the sequence, the ratio of two successive terms always approaches the Golden Ratio.

The Fibonacci numbers, like the "aspiration" of their ratios—the Golden Ratio—have some truly amazing properties. The list of mathematical relations involving Fibonacci numbers is literally endless. Here are just a handful of them.

"Squaring" Rectangles

If you sum up an odd number of products of successive Fibonacci numbers, like the three products $1 \times 1 + 1 \times 2 + 2 \times 3$, then the sum ($1 + 2 + 6 = 9$) is equal to the square of the last Fibonacci number you used in the products (in this case, $3^2 = 9$). To take another example, if we sum

up seven products, $1 \times 1 +$
$1 \times 2 + 2 \times 3 + 3 \times 5 + 5$
$\times 8 + 8 \times 13 + 13 \times 21 =$
441, the sum (441) is equal
to the square of the last
number used ($21^2 = 441$).
Similarly, summing up the
eleven products $1 \times 1 + 1$
$\times 2 + 2 \times 3 + 3 \times 5 + 5 \times$
$8 + 8 \times 13 + 13 \times 21 + 21$
$\times 34 + 34 \times 55 + 55 \times 89$
$+ 89 \times 144 = 144^2$. This
property can be represented
beautifully by a figure (Fig-

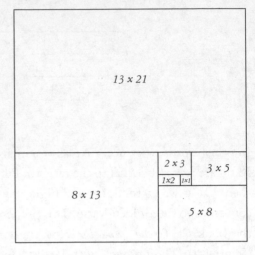

Figure 30

ure 30). Any odd number of rectangles with sides equal to successive
Fibonacci numbers fits precisely into a square. The figure shows an example with seven such rectangles.

Eleven Is the Sin

In the drama *The Piccolomini* by the German playwright and poet
Friedrich Schiller, astrologer Seni declares: "Elf ist die Sünde. Elfe über-
schreiten die zehn Gebote" ("Eleven is the sin. Eleven transgresses
the Ten Commandments"), expressing an opinion that dates back
to medieval times. The Fibonacci sequence, on the other hand, has a
property related to the number 11, which, far from being sinful, is
quite beautiful.

Suppose we sum up the first ten consecutive Fibonacci numbers:
$1 + 1 + 2 + 3 + 5 + 8 + 13 + 21 + 34 + 55 = 143$. This sum
is divisible evenly by 11 (143/11 = 13). The same is true for the sum of
any ten consecutive Fibonacci numbers. For example, $55 + 89 + 144 +$
$233 + 377 + 610 + 987 + 1{,}597 + 2{,}584 + 4{,}181 = 10{,}857$, and
10,857 is divisible by 11, 10,857/11 = 987. If you examine these two
examples, you discover something else. The sum of any ten consecutive
numbers is always equal to 11 times the seventh number. You can use
this property to amaze an audience by the speed with which you can add
any ten successive Fibonacci numbers.

Revenge of the Sexagesimal?

As you recall, for reasons that are not entirely clear, the ancient Babylonians used base 60 (the sexagesimal base) in their counting system. Although not related to the Babylonian number system, the number 60 happens to play a role in the Fibonacci sequence.

Fibonacci numbers become very large quite rapidly, because you always add two successive Fibonacci numbers to find the next one. In fact, we are quite lucky that rabbits don't live forever, or we would all be inundated with rabbits. While the fifth Fibonacci number is only 5, the 125th is already 59,425,114,757,512,643,212,875,125. Interestingly, the unit digit repeats itself with a periodicity of 60 (namely, after every 60 numbers). For example, the second number is 1, the sixty-second number is 4,052,739,537,881 (also ending in 1); the 122nd number, 14,028,366,653,498,915,298,923,761, also ends in 1; and so does the 182nd; and so on. Similarly, the fourteenth number is 377; the seventy-fourth number (sixty numbers farther along the sequence) 1,304,969,544,928,657 also ends in 7; and so on. This property was discovered in 1774 by the Italian-born French mathematician Joseph Louis Lagrange (1736–1813), who is responsible for many works in number theory and mechanics and who also studied the stability of the solar system. The last two digits (e.g., 01, 01, 02, 03, 05, 08, 13, 21 . . .) repeat in the sequence with a periodicity of 300 and the last three digits repeat with a periodicity of 1,500. In 1963 Stephen P. Geller used an IBM 1620 computer to show that the last four digits repeat every 15,000 times, the last five repeat every 150,000 times, and finally, after the computer ran for nearly three hours, a repetition of the last six digits appeared at the 1,500,000th Fibonacci number. Being unaware of the fact that a general theorem concerning the periodicity of the last digits could be proven, Geller commented: "There does not yet seem to be any way of guessing the next period, but perhaps a new program for the machine which will permit initialization at any point in the sequence for a test will cut down computer time enough so that more data can be gathered." Shortly thereafter, however, Israeli mathematician Dov Jarden pointed out that one can prove rigorously that for any number of last digits from three and up, the periodicity is simply fifteen times ten to a power that is one less than the number of digits (e.g., for seven digits it is 15×10^6, or 15 million).

Why $\frac{1}{89}$?

The properties of our universe, from the sizes of atoms to the sizes of galaxies, are determined by the values of a few numbers known as constants of nature. These constants include a measure of the strengths of all the basic forces—gravitational, electromagnetic, and two nuclear forces. The strength of the familiar electromagnetic force between two electrons, for example, is expressed in physics in terms of a constant known as the fine structure constant. The value of this constant, almost exactly $\frac{1}{137}$, has puzzled many generations of physicists. A joke made about the famous English physicist Paul Dirac (1902–1984), one of the founders of quantum mechanics, says that upon arrival to heaven he was allowed to ask God one question. His question was: "Why $\frac{1}{137}$?"

The Fibonacci sequence also contains one absolutely remarkable number—its eleventh number, 89. The value of $\frac{1}{89}$ in decimal representation is equal to: $0.01123595\ldots$ Suppose you arrange the Fibonacci numbers 1, 1, 2, 3, 5, 8, 13, 21, . . . as decimal fractions in the following way:

$$0.01$$
$$0.001$$
$$0.0002$$
$$0.00003$$
$$0.000005$$
$$0.0000008$$
$$0.00000013$$
$$0.000000021$$

. . .

In other words, the units digit in the first Fibonacci number is in the second decimal place, that of the second is in the third decimal place, and so on (the units digit of the nth Fibonacci number is in the $(n + 1)$th

decimal place). Now add all of those numbers up. In the preceding list
we would obtain 0.01123595 . . . , which is equal to ⅟₈₉.

Lightning Addition Trick

Some people can add numbers very quickly in their heads. The Fibonacci sequence allows a person to perform such lightning addition tricks without much effort. The sum of all the Fibonacci numbers from the first to the nth is simply equal to the $(n + 2)$th number minus 1. For example, the sum of the first ten numbers, $1 + 1 + 2 + 3 + 5 + 8 + 13 + 21 + 34 + 55 = 143$, is equal to the twelfth number (144) minus 1. The sum of the first seventy-eight numbers is equal to the eightieth number minus 1; and so on. Therefore, you can have someone write a long column of numbers starting with 1, 1, and continuing using the definition of the Fibonacci sequence (that each new number be the sum of the two previous ones). Tell this person to draw a line between some arbitrary two numbers in the column and you will be able, at a glance, to give the sum of all the numbers prior to the line. That sum will be equal to the second term after the line minus one.

Pythagorean Fibonaccis

Oddly enough, Fibonacci numbers can even be related to Pythagorean triples. The latter, as you recall, are triples of numbers that can serve as the lengths of the sides of a right-angled triangle (like the numbers 3, 4, 5). Take any four consecutive Fibonacci numbers, such as 1, 2, 3, 5. The product of the outer numbers, $1 \times 5 = 5$, twice the product of the inner terms, $2 \times 2 \times 3 = 12$, and the sum of the squares of the inner terms, $2^2 + 3^2 = 13$, give the three legs in the Pythagorean triple, 5, 12, 13 ($5^2 + 12^2 = 13^2$). But this is not all. Notice also that the third number, 13, is itself a Fibonacci number. This property was discovered by the mathematician Charles Raine.

Given the numerous wonders that the Fibonacci numbers hold in store (we shall soon encounter many more), it should come as no surprise that mathematicians looked for some efficient method for calculating these numbers, F_n, for any value of n. While in principle this is not a problem, since if we need the 100th number we simply have to add up the 98th and the 99th numbers, this still means that we first need to calculate all the numbers up to the 99th, which can be quite tedious. As the late comedian George Burns (in his book *How to Live to Be 100 or More*) once put it: "How do you live to be 100 or more? There are certain things you have to do. The most important one is you have to be sure to make it to 99."

In the middle of the nineteenth century, the French mathematician Jacques Phillipe Marie Binet (1786–1856) rediscovered a formula that was apparently known already in the eighteenth century to the most prolific mathematician in history, Leonard Euler (1707–1783), and to the French mathematician Abraham de Moivre (1667–1754). The formula allows you to find the value of any Fibonacci number, F_n, if its place in the sequence, n, is known. This Binet formula relies entirely on the Golden Ratio

$$F_n = \frac{1}{\sqrt{5}} \left[\left(\frac{1 + \sqrt{5}}{2} \right)^n - \left(\frac{1 - \sqrt{5}}{2} \right)^n \right].$$

At first glance, this is a formidably disconcerting formula, since it is not even obvious that upon substitution of various values of n it would produce whole numbers (which all the terms in the Fibonacci sequence are). Since we already know that the Fibonacci numbers are intimately related to the Golden Ratio, things start to look a little bit more reassuring when we realize that the first term inside the brackets is, in fact, simply the Golden Ratio raised to the nth power, ϕ^n, and the second is $(-1/\phi)^n$. (Recall from earlier that the negative solution of the quadratic equation defining ϕ is equal to $-1/\phi$.) Using a simple scientific pocket calculator you can test for a few values of n that Binet's formula is indeed giving the Fibonacci numbers correctly. For relatively large values of n, the second term in the brackets above becomes very small, and you can simply take F_n to be the closest whole number to $\frac{\phi^n}{\sqrt{5}}$. For example, for $n = 10$, $\frac{\phi^n}{\sqrt{5}}$ is equal to 55.0036, and the tenth Fibonacci number is 55.

Just as an amusement, you may wonder if there is a Fibonacci number with precisely 666 digits. Mathematician and author Clifford A. Pickover calls numbers associated with 666 "apocalyptic." He found that the 3,184th Fibonacci number has 666 digits.

Once discovered, Fibonacci numbers seemed to start popping up everywhere in nature. A few fascinating examples are provided by botany.

AS THE SUNFLOWER TURNS ON HER GOD

The leaves along a twig of a plant or the stems along a branch tend to grow in positions that would optimize their exposure to sun, rain, and air. As a vertical stem grows, it produces leaves at quite regular spacings. However, the leaves do not grow directly one above the other, because this would shield the lower leaves from the moisture and sunlight they need. Rather, the passage from one leaf to the next (or from one stem to the next along a branch) is characterized by a screw-type displacement around the stem (as in Figure 31). Similar arrangements of repeating units can be found in the scales of a pinecone or the seeds of a sunflower. This phenomenon is called *phyllotaxis* ("leaf arrangement" in Greek), a word coined in 1754 by the Swiss naturalist Charles Bonnet (1720–1793). For example, in basswoods leaves occur generally on two opposite sides (corresponding to half a turn around the stem), which is known as a ½ phyllotactic ratio. In other plants, such as the hazel, blackberry, and beech, passing from one leaf to the next involves one-third of a turn (⅓ phyllotactic ratio). Similarly, the apple, the coast live oak, and the apricot have leaves every ⅖ of a turn, and the pear and the weeping willow have them every ⅜ of a turn. Figure 31 illustrates a case where it took three complete turns

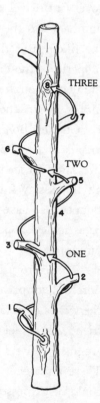

Figure 31

to pass through eight stems (a phyllotactic ratio of ⅜). You'll notice that all the fractions that are observed are ratios of alternate members of the Fibonacci sequence.

The fact that leaves of plants follow certain patterns was first noted in antiquity by Theophrastus (ca. 372 B.C.–ca. 287 B.C.) in *Enquiry into Plants*. He remarks: "those that have flat leaves have them in a regular series." Pliny the Elder (A.D. 23–79) made a similar observation in his monumental *Natural History,* where he talks about "regular intervals" between leaves "arranged circularly around the branches." The study of phyllotaxis did not go much beyond these early, qualitative observations until the fifteenth century, when Leonardo da Vinci (1452–1519) added a quantitative element to the description of leaf arrangements by noting that the leaves were arranged in spiral patterns, with cycles of five (corresponding to an angle of ⅖ of a turn). The first person to discover (intuitively) the relation between phyllotaxis and the Fibonacci numbers was the astronomer Johannes Kepler. Kepler wrote: "It is in the likeness of this self-developing series [referring to the recursive property of the Fibonacci sequence] that the faculty of propagation is, in my opinion, formed; and so in a flower the authentic flag of this faculty is shown, the pentagon."

Charles Bonnet initiated serious studies in observational phyllotaxis. In his 1754 book *Recherches sur l'Usage des Feuilles dans les Plantes* (Research on the use of leaves in plants) he gives a clear description of ⅖ phyllotaxis. While working with the mathematician G. L. Calandrini, Bonnet may have also discovered that sets of spiral rows (now known as parastichies) appear in some plants, like the scales of a fir cone or a pineapple.

The history of truly *mathematical* phyllotaxis (as opposed to the purely descriptive approaches) begins in the nineteenth century with the works of botanist Karl Friedric Schimper (published in 1830), his friend Alexander Braun (published in 1835), and the crystallographer Auguste Bravais and his botanist brother Louis (published in 1837). These researchers discovered the general rule that phyllotactic ratios could be expressed by ratios of terms of the Fibonacci series (like ⅖; ⅜) and also noted the appearance of consecutive Fibonacci numbers in the parastichies of pinecones and pineapples.

Pineapples indeed provide a truly beautiful manifestation of a Fibonacci-based phyllotaxis (Figure 32). Each hexagonal scale on the surface of the pineapple is a part of three different spirals. In the figure you can see one of eight parallel rows sloping gently from lower left to upper right, one of thirteen parallel rows that slope more steeply from lower right to upper left, and one of twenty-one parallel rows that are very steep (from lower left to upper right). Most pineapples have five, eight, thirteen, or twenty-one spirals of increasing steepness on their surface. All of these are Fibonacci numbers.

Figure 32   Figure 33

How do plants know to arrange their leaves in these Fibonacci patterns? The growth of the plant takes place at the tip of the stem (called the meristem), which has a conical shape (being thinnest at the tip). Leaves that are farther down from the tip (namely, which grew earlier) tend to be radially farther out from the stem's center when viewed from the top (because the stem is thicker there). Figure 33 shows such a view of the stem from the top, where the leaves are numbered according to their order of appearance. The leaf numbered 0, which appeared first, is by now the farthest down from the meristem and the farthest out from the stem's center. Botanist A. H. Church in his 1904 book *On the Relation of Phyllotaxis to Mechanical Laws* first emphasized the importance of this type of representation for the understanding of phyllotaxis. What we find (by imagining a curve that connects leaves 0 to 5 in Figure 33) is that successive leaves sit along a tightly wound spiral, known as the generative spiral. The important quantity that characterizes the location of the leaves is the angle between the lines connecting the stem's center with successive leaves. One of the discoveries of the Bravais brothers in 1837 was that new leaves advance roughly by the same angle around the circle and that this angle (known as the divergence angle) is usually close to 137.5 degrees. Are you shocked to hear that this value is determined by the Golden Ratio? The angle that divides a complete turn in a Golden Ratio is 360°/φ = 222.5 degrees. Since this is more than half a circle (180 degrees), we should measure it going in the opposite direction around the circle. In other words, we should subtract 222.5 from 360, giving us the observed angle of 137.5 degrees (sometimes called the Golden Angle).

In a pioneering work in 1907, German mathematician G. van Iterson showed that if you closely pack successive points separated by 137.5 degrees on tightly wound spirals, then the eye would pick out one family of spiral patterns winding clockwise and one counterclockwise. The numbers of spirals in the two families tend to be consecutive Fibonacci numbers, since the ratio of such numbers approaches the Golden Ratio.

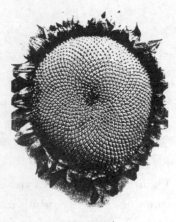

Figure 34

Such counterwinding spirals are most spectacularly exhibited by the arrangement of the florets in sunflowers. When you look on the head of a sunflower (Figure 34), you will notice both clockwise and counterclockwise spiral patterns formed by the florets. Clearly the florets grow in a way that affords the most efficient sharing of horizontal space. The numbers of these spirals usually depend on the size of the sunflower. Most commonly there are thirty-four spirals going one way and fifty-five the other, but sunflowers with ratios of numbers of spirals of 89/55, 144/89, and even (at least one; reported by a Vermont couple to the *Scientific American* in 1951) 233/144 have been seen. All of these are, of course, ratios of adjacent Fibonacci numbers. In the largest sunflowers, the structure stretches from one pair of consecutive Fibonacci numbers to the next higher, when we move from the center to the periphery.

The petal counts and petal arrangements of some flowers also harbor Fibonacci numbers and Golden Ratio connections. Many people have relied (at least symbolically) at some point in their lives on the numbers of petals of daisies to satisfy their curiosity about the intriguing question: "She loves me, she loves me not." Most field daisies have thirteen, twenty-one, or thirty-four petals, all Fibonacci numbers. (Wouldn't it be nice to know in advance if the daisy has an even or odd number of petals?) The number of petals simply reflects the number of spirals in one family.

The beautifully symmetric arrangement of the petals of roses is also based on the Golden Ratio. If you dissect a rose (petal by petal), you will discover the positions of its tightly packed petals. Figure 35 presents a schematic in which the petals have been numbered. The angles defining the positions (in fractions of a full turn) of the petals are the fractional part of simple multiples of φ. Petal 1 is 0.618 (the fractional part of 1 × φ) of a turn from petal 0, petal 2 is 0.236 (the fractional part of 2 × φ) of a turn from petal 1, and so on.

This description shows that the 2,300-year-old puzzle of the origins of phyllotaxis reduces to the basic question: Why are successive leaves separated by the Golden Angle of 137.5 degrees? The attempts to answer this question come in two flavors: theories that concentrate on the geometry of the configuration, and on simple mathematical rules that can generate this geometry; and models that suggest

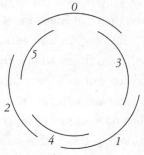

Figure 35

an actual dynamical cause for the observed behavior. Landmark works of the first type (e.g., by mathematicians Harold S. M. Coxeter and I. Adler and by crystallographer N. Rivier) show that buds which are placed along the generative spiral separated by the Golden Angle are close-packed most efficiently. This is easy to understand. If the divergence angle was, let's say, 120 degrees (which is 360/3) or any other rational multiple of 360 degrees, then the leaves would have aligned radially (along three lines in the case of 120 degrees), leaving large spaces in between. On the other hand, a divergence angle like the Golden Angle (which is an irrational multiple of 360 degrees) ensures that buds do not line up along any specific radial direction and they fill the spaces efficiently. The Golden Angle proves to be even better than other irrational multiples of 360 degrees because the Golden Ratio is the most irrational of all irrational numbers in the following sense. Recall that the Golden Ratio is equal to a continued fraction composed entirely of 1s. That continued fraction converges more slowly than any other continued fraction. In other words, the Golden Ratio is farther away from being expressible as a fraction than any other irrational number.

In a paper that appeared in 1984 in *Journal de Physique,* a team of scientists led by N. Rivier from the Université de Provence in Marseille, France, used a simple mathematical algorithm to show that when a growth angle equal to the Golden Angle is used, structures that closely resemble real sunflowers are obtained. (See Figure 36.) Rivier and his collaborators suggested that this provided an answer to the question that had been posed in the classical work of biologist Sir D'Arcy Wentworth Thompson. In his monumental book *On Growth and Form* (first published in 1917 and revised in 1942), Thompson wonders: ". . . and not the least curious feature of the case [phyllotaxis] is the limited, even the small number of possible arrangements which we observe and recognize." Rivier's team found that the requirements of *homogeneity* (that the structure is the same everywhere) and of *self-similarity* (that when one examines the structure on different scales from small to large, it looks precisely the same) limit drastically the number of possible structures. These two properties may be sufficient to explain the preponderance of Fibonacci numbers and the Golden Ratio in phyllotaxis, but they still do not offer any physical cause.

The best clues for a possible dynamical cause of phyllotaxis came not from botany but from experiments in physics by L. S. Levitov (in 1991) and by Stephane Douady and Yves Couder (in 1992 to 1996). The experiment by Douady and Couder is particularly fascinating. They held a dish full of silicone oil in a magnetic field that was stronger near the dish's edge than at the center. Drops of a magnetic fluid, which act like tiny bar magnets, were dropped periodically at the center of the dish. The

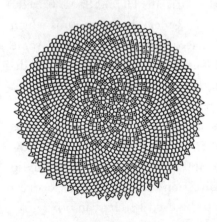

Figure 36

tiny magnets repelled each other and were pushed radially by the magnetic field gradient. Douady and Couder found patterns that oscillated about, but generally converged to, a spiral on which the Golden Angle separated successive drops. Physical systems usually settle into states that minimize the energy. The suggestion is therefore that phyllotaxis simply represents a state of minimal energy for a system of mutually repelling buds. Other models, in which leaves appear at the points of the highest concentration of some nutrient, also tend to produce separations equal to the Golden Angle.

I hope that the next time you eat a pineapple, send a red rose to a loved one, or admire van Gogh's sunflower paintings, you will remember that the growth pattern of these plants embodies this wonderful number we call the Golden Ratio. Realize, however, that plant growth also depends on factors other than optimal spacing. Consequently, the phyllotaxis rules I have described cannot be taken as applying to all circumstances, like a law of nature. Rather, in the words of the famous Canadian mathematician Coxeter, they are "only a fascinatingly prevalent *tendency.*"

Botany is not the only place in nature where the Golden Ratio and Fibonacci numbers can be found. They appear in phenomena covering a range in sizes from the microscopic to that of giant galaxies. Often that appearance takes the form of a magnificent spiral.

ALTHOUGH CHANGED, I RISE AGAIN THE SAME

No family in the history of mathematics has produced as many celebrated mathematicians (thirteen in total!) as did the Bernoulli family. Disconcerted by the Spanish Fury (the ravaging riot in the Netherlands by Spanish soldiers), the family fled to Basel, Switzerland, from the Catholic Spanish Netherlands. Three members of the family, the brothers Jacques (1654–1705) and Jeanne (1667–1748), and the latter's second son, Daniel (1700–1782), stood out head and shoulders above the rest. Strangely, the Bernoullis were almost equally famous for their bitter interfamilial rivalries as they were for their numerous mathematical achievements. In one case, the exchanges between Jacques and Jeanne became particularly acrimonious. The feud was sparked by a dispute over a solution to a famous problem in mechanics. The problem, known as the brachistochrone (from the Greek *brachistos,* "shortest," and *chronos,* "time"), was to find the curve along which a particle acted on by the force of gravity will pass in the shortest time from one point to another. The two brothers proposed the same solution independently, but Jeanne's derivation was incorrect, and he later attempted to present Jacques' derivation as his own. The sad consequence of this chain of events was that Jeanne became a professor in Groningen and did not set foot in Basel until his brother's death.

Jacques Bernoulli's association with the Golden Ratio comes through another famous curve. He devoted a treatise entitled *Spira Mirabilis* (Wonderful spiral) to a particular type of spiral shape. Jacques was so impressed with the beauty of the curve known as a logarithmic spiral (Figure 37; the name was derived from the way in which the radius grows as we move around the curve clockwise) that he asked that this shape, and the motto he assigned to it: "Eadem mutato resurgo" (although changed, I rise again the same), be engraved on his tombstone.

The motto describes a fundamental property unique to the logarithmic spiral—it does not alter its shape as its size increases. This feature is known as self-similarity. Fascinated by this property, Jacques wrote that the logarithmic spiral "may be used as a symbol, either of fortitude and constancy in adversity, or of the human body, which after all its changes, even after death, will be restored to its exact and perfect self."

Figure 37

If you think about it for a moment, this is precisely the property required for many growth phenomena in nature. For example, as the mollusk inside the shell of the chambered nautilus (Figure 4) grows in size, it constructs larger and larger chambers, sealing off the smaller unused ones. Each increment in the length of the shell is accompanied by a proportional increase in its radius, so that the shape remains un-

changed. Consequently, the nautilus sees an identical "home" throughout its lifetime, and it does not need, for example, to adjust its balance
as it matures. The latter property applies also to rams, the horns of
which are also in the shape of logarithmic spirals (although they do not
lie in a plane), and to the curve of elephants' tusks. Increasing by ac-
cumulation from within itself, the logarithmic spiral grows wider,
with the distance between its "coils" increasing, as it moves away from
the source, known as the pole. Specifically, turning by equal angles increases the distance from the pole by equal ratios. If we were, with the
aid of a microscope, to enlarge the coils that are invisible to the naked
eye to the size of Figure 37, they would fit precisely on the larger spiral. This property distinguishes the logarithmic spiral from another
common spiral known as the Archimedean spiral, after the famous
Greek mathematician Archimedes (ca. 287–212 B.C.), who described
it extensively in his book *On Spirals*. We can see an Archimedean spiral in the side of a roll of paper towels or a rope coiled on the floor. In
this type of spiral, the distance between successive coils remains always
the same. As a result of a mistake that surely would have caused
Jacques Bernoulli much grief, the mason who prepared Bernoulli's
tombstone engraved on it an Archimedean rather than a logarithmic
spiral.

Nature loves logarithmic spirals. From sunflowers, seashells, and
whirlpools, to hurricanes and giant spiral galaxies, it seems that nature
chose this marvelous shape as its favorite "ornament." The constant
shape of the logarithmic spiral on all size scales reveals itself beautifully
in nature in the shapes of minuscule fossils or unicellular organisms
known as foraminifera. Although the spiral shells in this case are composite structures (and not one continuous tube), X-ray images of the internal structure of these fossils show that the shape of the logarithmic
spiral remained essentially unchanged for millions of years.

In his classic book *The Curves of Life* (1914), English author and editor Theodore Andrea Cook gives numerous examples of the appearance
of spirals (not just logarithmic) in nature and art. He discusses spirals in
things as diverse as climbing plants, the human body, staircases, and
Maori tattoos. In explaining his motivation for writing the book, Cook
writes: "for the existence of these chapters upon spiral formations no

other apology is needed than the interest and beauty of an investigation."

Artists have also not failed to see the beauty of logarithmic spirals. In Leonardo da Vinci's study for the mythological subject "Leda and the Swan," for example, he draws the hair arranged in the shape of a nearly logarithmic spiral (Figure 38). Leonardo repeats this shape many times in his study of spirals in clouds and water in the impressive series of sketches for the "Deluge." In that work, he combined his scientific observations of frightening floods with the allegorical aspects of destructive forces descended from heaven. Describing the violent flow of water Leonardo wrote: "The sudden waters rush into the pond that contains them, striking the various obstacles with swirling eddies. . . . The momentum of the circular waves flying from the point of impact hurls them in the way of other circular waves moving in the opposite direction."

Figure 38   Figure 39

Twentieth-century designer and illustrator Edward B. Edwards developed hundreds of decorative designs based on the logarithmic spiral; many can be seen in his book *Pattern and Design with Dynamic Symmetry* (an example is shown in Figure 39).

The logarithmic spiral and the Golden Ratio go hand in hand. Examine again the series of nested Golden Rectangles obtained when you snip off squares from a Golden Rectangle (Figure 40; we encountered this property already in Chapter 4). If you connect the successive points where these "whirling squares" divide the sides in Golden Ratios, you obtain a logarithmic spiral that coils inward toward the pole (the point given by the intersection of the diagonals in Figure 25, which was called fancifully "the eye of God").

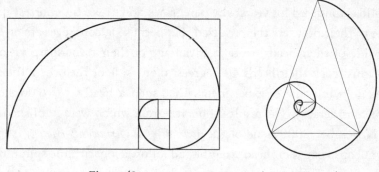

Figure 40   Figure 41

You can also obtain a logarithmic spiral from a Golden Triangle. We saw in Chapter 4 that if you start from a Golden Triangle (an isosceles triangle in which the side is in Golden Ratio to the base) and bisect a base angle, you get a smaller Golden Triangle. If you continue the process of bisecting the base angle ad infinitum, you will generate a series of whirling triangles. Connecting the vertices of the Golden Triangles in the progression will trace a logarithmic spiral (Figure 41).

The logarithmic spiral is also known as the *equiangular spiral*. This name was coined in 1638 by the French mathematician and philosopher René Descartes (1596–1650), after whom we name the numbers used to locate a point in the plane (with respect to two axes)—Cartesian coordinates. The name "equiangular" reflects another unique property of the logarithmic spiral. If you draw a straight line from the pole to any point on the

Figure 42

curve, it cuts the curve at precisely the same angle (Figure 42). Falcons use this property when attacking their prey. Peregrine falcons are some of the fastest birds on Earth, plummeting toward their targets at speeds of up to two hundred miles per hour. But they could fly even faster if they would just fly straight instead of following a spiral trajectory to their victims. Biologist Vance A. Tucker of Duke University in North Carolina wondered for years why peregrines don't take the shortest distance to their prey. He then realized that because falcons' eyes are on either side of their heads, to take advantage of their razor-sharp vision, they must cock their heads 40 degrees to one side or the other. Tucker found in wind-tunnel experiments that such a head tip would slow them considerably. The results of his research, which were published in the November 2000 issue of the *Journal of Experimental Biology,* show that falcons keep their head straight and follow a logarithmic spiral. Because of the spiral's equiangular property, this path allows them to keep their target in view while maximizing speeds.

The amazing thing is that the same spiral shape that is found in the unicellular foraminifera and in the sunflower and that guides the flight of a falcon can also be found in those "systems of stars gathered together in a common plane, like those of the Milky Way" which philosopher Immanuel Kant (1724–1804) speculated about long before they were actually observed (Figure 43). These became known as island universes— giant galaxies containing hundreds of billions of stars like our Sun. Observations conducted with the Hubble Space

Figure 43

Telescope revealed that there are some one hundred billion galaxies in our observable universe, many of which are spiral galaxies. You can hardly think of a better manifestation of the grand vision expressed by English poet, painter, and mystic William Blake (1757–1827), when he wrote:

To see a World in a Grain of Sand,
And a Heaven in a Wild Flower,
Hold Infinity in the Palm of your hand,
And Eternity in an hour.

Why do so many galaxies exhibit a spiral pattern? Spiral galaxies like
our own Milky Way have a relatively thin disk (like a pancake) composed of gas, dust (miniature grains), and stars. The entire galactic disk
is rotating about the galactic center. In the vicinity of the Sun, for example, the orbital speed around the Milky Way's center is about 140
miles per second, and it takes material about 225 million years to complete one revolution. At other distances from the center the speed is different—higher closer to the center, lower at greater distances—that is,
galactic disks do not rotate like a solid compact disk but rather rotate
differentially. Seen face on, spiral galaxies show spiral arms originating
close to the galactic center and extending outward throughout much of
the disk (as in the "Whirlpool Galaxy," Figure 43). The spiral arms are
the part of the galactic disk where many young stars are being born.
Since young stars are also the brightest, we can see the spiral structure
of other galaxies from afar. The basic question that astrophysicists had
to answer is: How do the spiral arms retain their shape over long periods of time? Because the inner parts of the disk rotate faster than the
outer parts, any large-scale pattern that is somehow "attached" to the
disk material (e.g., the stars) cannot survive for long. A spiral structure
tied always to the same collection of stars and gas clouds would in-
evitably wind up, contrary to observations. The explanation for the
longevity of the spiral arms relies on *density waves*—waves of gas com-
pression sweeping through the galactic disk—squeezing gas clouds
along the way and triggering the formation of new stars. The spiral pattern that we observe simply marks the denser-than-average parts of the
disk and its newborn stars. The pattern is therefore created repeatedly
without winding up. The situation is similar to that observed near a
lane closed for repairs by a work crew on a major highway. The density
of cars in the vicinity of the closed stretch is higher because cars have to
slow down there. If you take a long-exposure photograph of the highway from above, you will record the high-traffic density near the place

where repairs are being undertaken. Just as the traffic density wave is not associated with any particular set of cars, the spiral-arms pattern is not tied to any particular piece of disk material. Another similarity is in the fact that the density wave itself moves through the disk more slowly than the motion of the stars and the gas, just as the speed at which the repair work proceeds along the highway is typically much slower than the unperturbed speed of individual cars.

The agent that deflects the motion of the stars and the gas clouds and generates the spiral density wave (analogous to the repair crew that deflects the moving cars to fewer lanes) is the gravitational force resulting from the fact that the distribution of matter in the galaxy is not perfectly symmetric. For example, a set of oval orbits around the center (Figure 44a) in which each orbit is perturbed (rotated) slightly by an amount that changes with distance from the center results in a spiral pattern (Figure 44b).

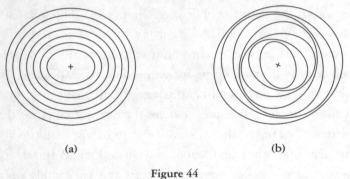

(a)   (b)

Figure 44

Actually, we should be quite happy that the force of gravity behaves in our universe the way it does. According to Newton's universal law of gravitation, every mass attracts every other mass, and the force of attraction decreases as the masses get farther apart. In particular, doubling the distance weakens the force by a factor of four (the force decreases as the inverse of the square of the distance). Newton's laws of motion show that as a result of this dependence on the distance, the orbits of the planets around the Sun are in the shapes of ellipses. Imagine what would have happened had we lived in a universe in which gravity had decreased by a factor of eight (instead of four) upon doubling of the distance (a force decreasing as the inverse of the cube of the distance). In such a universe, Newton's laws predict that one possible orbit of the planets is a logarithmic spiral. In other words, Earth would have spiraled into the Sun or rushed off into space.

Leonardo Fibonacci, who initiated all of this frenzy of mathematical activity, is far from forgotten today. In today's Pisa, a statue of Fibonacci constructed in the nineteenth century stands in the Scotto Garden on the grounds of the Sangallo Fortress, next to a street named after Fibonacci, which runs along the south side of the Arno River.

Since 1963 the Fibonacci Association has published a journal entitled the *Fibonacci Quarterly*. The group was formed by mathematicians Verner Emil Hoggatt (1921–1981) and Brother Alfred Brousseau (1907–1988) "in order to exchange ideas and stimulate research in Fibonacci numbers and related topics." Perhaps against the odds, the *Fibonacci Quarterly* has since grown into a well-recognized journal in number theory. As Brother Brousseau humorously put it: "We got a group of people together in 1963, and just like a bunch of nuts, we started a mathematics magazine." The Tenth International Conference on Fibonacci Numbers and Their Applications is planned for June 24–28, 2002, at Northern Arizona University in Flagstaff, Arizona.

All of this is but a small tribute to the man who used rabbits to discover a world-embracing mathematical concept. As important as Fibonacci's contribution was, however, the story of the Golden Ratio did not end in the thirteenth century; fascinating developments were still to come in Renaissance Europe.

6

THE DIVINE
PROPORTION

*The quest for our origin is that sweet fruit's juice which maintains
satisfaction in the minds of the philosophers.*
—LUCA PACIOLI (1445–1517)

Few famous painters in history have also been gifted mathematicians. However, when we speak of a "Renaissance man," we mean a person who exemplifies the Renaissance ideal of wide-ranging culture and learning. Accordingly, three of the best-known Renaissance painters, the Italians Piero della Francesca (ca. 1412–1492) and Leonardo da Vinci and the German Albrecht Dürer, also made interesting contributions to mathematics. Not surprisingly perhaps, the mathematical investigations of all three painters were related to the Golden Ratio.

The most active mathematician of this illustrious trio of artists was Piero della Francesca. The writings of Antonio Maria Graziani (the brother-in-law of Piero's great-grandchild), who purchased Piero's house, indicate that the artist was born in 1412 in Borgo San Sepolcro (today Sansepolcro) in central Italy. His father, Benedetto, was a prosperous tanner and shoemaker. Little else is known about Piero's very early life, but newly discovered documents show that he spent some time before 1431 as an apprentice in the workshop of the painter Antonio D'Anghiari (by whom no works have survived). By the late 1430s Piero had moved to Florence, where he started to work with the artist Domenico Veneziano. In Florence, the young painter was exposed to the works of such early Renaissance painters as Fra Angelico and Masaccio and to the sculptures of Donatello. He was particularly impressed with the serenity of the religious works of Fra Angelico, and his own style, in terms of application of color and light, reflected this influence. Later in life, every phase in Piero's work was characterized by a burst of activity, in a variety of places including Rimini, Arezzo, and Rome. The figures that Piero painted either have an architectural solidity about them, as in the "Flagellation of Christ" (currently in the Galleria Nationale delle Marche in Urbino; Figure 45), or they seem like natural exten-

Figure 45   Figure 46

sions of the background, as in "The Baptism" (currently in the National Gallery, London; Figure 46).

In the *Lives of the Most Eminent Painters, Sculptors, and Architects,* the first art historian, Giorgio Vasari (1511–1574), writes that Piero demonstrated great mathematical ability since early youth, and he attributes to him "many" mathematical treatises. Some of these were written when the painter, because of his old age, could no longer practice art. In the dedicatory letter to Duke Guidobaldo of Urbino, Piero says about one of his books that it was composed "in order that his wits might not go torpid with disuse." Three of Piero's mathematical works have survived: *De Prospectiva pingendi* (On perspective in painting), *Libellus de Quinque Corporibus Regularibus* (Short book on the five regular solids), and *Trattato d'Abaco* (Treatise on the abacus).

Piero's *On Perspective* (written in the mid-1470s to 1480s) contains numerous references to Euclid's *Elements* and *Optics,* since he was determined to demonstrate that the technique for achieving perspective in a painting relies firmly on the scientific basis for visual perception. In his own paintings, perspective provides a spacious container that is in complete consonance with the geometrical properties of the figures within. In fact, to Piero, painting itself was primarily "the demonstration in a plane of bodies in diminishing or increasing size." This attitude is manifested magnificently in the "Flagellation" (Figures 45 and 47), which is one of the few Renaissance paintings with a very meticulously determined perspectival construction. As modern-day artist David Hockney puts it in his 2001 book *Secret Knowledge,* Piero paints "the way he knows the figures to be, not the way he sees them."

Figure 47

With the occasion of the 500th anniversary of Piero's death, re-searchers Laura Geatti of the University of Rome and Luciano Fortunati of the National Research Council in Pisa performed a detailed, computer-aided analysis of the "Flagellation." They digitized the entire image, determined the coordinates of all the points, measured all the distances, and conducted a complete perspectival analysis using alge-braic calculations. Doing this allowed them to determine the precise lo-cation of the "vanishing point," at which all lines receding directly from the viewer converge (Figure 47), that Piero used to achieve the paint-ing's impressive "depth."

Piero's lucid book on perspective became the standard manual for artists who attempted to paint plane figures and solids, and the less mathematical (and more accessible) parts of the treatise were incorpo-rated into most subsequent works on perspective. Vasari testifies that due to Piero's strong mathematical background, "he understood better than anyone else all the curves in the regular bodies and we are indebted to him for the light shed on that subject." An example of Piero's careful analysis of how to draw a pentagon in perspective is shown in Figure 48.

In both the *Treatise on the Abacus* and the *Five Regular Solids,* Piero

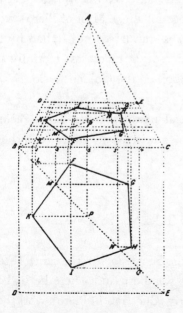

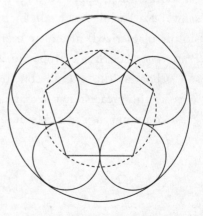

Figure 48   Figure 49

presents a wide range of problems (and their solutions) that involve the pentagon and the five Platonic solids. He calculates the lengths of sides and diagonals as well as areas and volumes. Many of the solutions involve the Golden Ratio, and some of Piero's techniques represent innovative thinking and originality.

Like Fibonacci before him, Piero wrote the *Treatise on the Abacus* mainly to provide the merchants of his day with arithmetic recipes and geometrical rules. In a commercial world that had neither a unique system of weights and measures nor even agreed-upon shapes or sizes of containers, the ability to calculate volumes of figures was an absolute must. However, Piero's mathematical curiosity carried him well beyond the subjects that had simple everyday applications. Accordingly, we find in his books "useless" problems, such as calculating the side of an octahedron inscribed inside a cube or calculating the diameter of the five small circles inscribed inside a circle of a known diameter (Figure 49). The solution of the latter problem involves a pentagon and, therefore, the Golden Ratio.

Much of Piero's algebraic work was incorporated into a book published by Luca Pacioli (1445–1517), entitled *Summa de arithmetica, geometria, proportioni et proportionalita* (The collected knowledge of arithmetic, geometry, proportion and proportionality). Most of Piero's work on solids, which appeared in Latin, was translated into Italian by the same Luca Pacioli and again incorporated (or, many less tactfully say, simply plagiarized) into his famous book on the Golden Ratio: *Divina Proportione* (The divine proportion).

Who was this highly controversial mathematician Luca Pacioli? Was he the greatest mathematical plagiarist of all times or rather a great communicator of mathematics?

UNSUNG HERO OF THE RENAISSANCE?

Luca Pacioli was born in 1445 in Borgo San Sepolcro (the same Tuscan town in which Piero della Francesca was born and where he had his workshop). In fact, Pacioli had his early education in Piero's workshop. However, unlike other students who displayed skill in the art of painting, and some, like Pietro Perugino, were destined to become great painters themselves, he showed greater promise in mathematics. Piero and Pacioli were closely associated later in life, as manifested by the fact that Piero included a portrait of Pacioli, as St. Peter Martyr, in a painting of "Madonna and Child with Saints and Angels." Pacioli moved to Venice at a relatively young age and became the tutor of the three sons of a wealthy merchant. In Venice he continued his mathematical education (under the mathematician Domenico Bragadino) and wrote his first textbook on arithmetic.

In the 1470s, Pacioli studied theology and was ordained as a Franciscan friar. Consequently, he is customarily referred to as Fra Luca Pacioli. In the following years, he traveled extensively, teaching mathematics at the universities of Perugia, Zara, Naples, and Rome. During this period he may have also tutored for some time Guidobaldo of Montefeltro, who was to become the Duke of Urbino in 1482.

In what may be the best portrait of a mathematician ever produced, Jacopo de' Barbari (1440–1515) depicts Luca Pacioli giving a lesson in

Figure 50

geometry to a pupil (Figure 50; the painting is currently in the Galleria Nazionale Di Capodimonte in Naples). One of the Platonic solids, a dodecahedron, is seen on the right resting on top of Pacioli's book *Summa*. Pacioli himself, dressed in his friar robes and almost resembling a geometrical solid, is shown copying a diagram from volume XIII of Euclid's *Elements*. A transparent polyhedron known as a rhombicubocta-hedron (one of the Archimedean Solids, with twenty-six faces of which eighteen are squares and eight equilateral triangles), half filled with water and hanging in mid-air, symbolizes the purity and timelessness of mathematics. The artist has captured the reflections and refractions from this glass polyhedron with extraordinary skill. The identity of the second person in the painting has been the subject of some debate. One of the suggestions is that the student is Duke Guidobaldo. British mathematician Nick MacKinnon raised an interesting possibility in 1993. In a well-researched article entitled "The Portrait of Fra Luca Pacioli," which appeared in the *Mathematical Gazette*, MacKinnon suggests that the figure is that of the famous German painter Albrecht Dürer, who had great interest in geometry and perspective (and to whose relationship with Pacioli we shall return later in this chapter). The face of the student does in fact bear a striking resemblance to Dürer's self-portrait.

Pacioli returned to Borgo San Sepolcro in 1489, after having been granted some special privileges by the Pope, only to encounter jealousy from the existing religious establishment. For about two years he was even banned from teaching. In 1494, Pacioli went to Venice to publish his *Summa*, which he dedicated to Duke Guidobaldo. Encyclopedic in nature and scope (some 600 pages), the *Summa* compiled the mathematical knowledge of the time in arithmetic, algebra, geometry, and trigonometry. In this book, Pacioli borrows freely (usually with an appropriate acknowledgment) problems on the icosahedron and dodeca-hedron from Piero's *Trattato* and problems in algebra and geometry from Fibonacci and others. Identifying Fibonacci as his main source, Pacioli states that when no other is quoted, the work belongs to Leonardus Pisanus. An interesting part of the *Summa* is on double-entry accounting, a method of record keeping that lets you track where money comes from and where it goes. While Pacioli did not invent this system but merely summarized the practices of Venetian merchants during the Renaissance, this is considered to be the first published book on accounting. Pacioli's desire to "give the trader without delay information as to his assets and liabilities" thus gained him the title "Father of Accounting," and accountants from all over the world celebrated in 1994 (in Sansepolcro, as the town is now known) the 500th anniversary of the *Summa*.

In 1480, Ludovico Sforza became effectively the Duke of Milan. In fact, he was only the regent of the real seven-year-old duke, following an episode of political intrigue and murder. Determined to make his court a home for scholars and artists, Ludovico invited Leonardo da Vinci in 1482 as a "painter and engineer of the duke." Leonardo had considerable interest in geometry, especially for its practical applications in mechanics. In his words: "Mechanics is the paradise of the mathematical sciences, because by means of it one comes to the fruits of mathematics." Consequently, Leonardo was probably the one who induced the duke to invite Pacioli to join the court, as a teacher of mathematics, in 1496. Undoubtedly, Leonardo learned some of his geometry from Pacioli, while he infused in the latter a greater appreciation for art.

During his stay in Milan, Pacioli completed work on his three-volume treatise *Divina Proportione* (The divine proportion), which was eventually published in Venice in 1509. The first volume, *Compendio de Divina Proportione* (Compendium of the divine proportion), contains a detailed summary of the properties of the Golden Ratio (which Pacioli refers to as the "Divine Proportion") and a study of Platonic solids and other polyhedra. On the first page of *The Divine Proportion* Pacioli declares somewhat grandiloquently that this is: "A work necessary for all the clear-sighted and inquiring human minds, in which everyone who loves to study philosophy, perspective, painting, sculpture, architecture, music and other mathematical disciplines will find a very delicate, subtle and admirable teaching and will delight in diverse questions touching upon a very secret science."

Pacioli dedicated the first volume of *The Divine Proportion* to Ludovico Sforza, and in the fifth chapter he lists five reasons why he believes that the appropriate name for the Golden Ratio should be *The Divine Proportion.*

1. "That it is one only and not more." Pacioli compares the unique value of the Golden Ratio to the fact that unity "is the supreme epithet of God himself."

2. Pacioli finds a similarity between the fact that the definition of the Golden Ratio involves precisely three lengths *(AC, CB,* and *AB* in Figure 24) and the existence of a Holy Trinity, of Father, Son, and Holy Ghost.

3. To Pacioli, the incomprehensibility of God and the fact that the Golden Ratio is an irrational number are equivalent. In his own words: "Just like God cannot be properly defined, nor can be understood through words, likewise our proportion cannot be ever designated by intelligible numbers, nor can it be expressed by any rational quantity, but always remains concealed and secret, and is called irrational by the mathematicians."

4. Pacioli compares the omnipresence and invariability of God to the self-similarity associated with the Golden Ratio—that its value is always the same and does not depend on the length of the line being divided or the size of the pentagon in which ratios of lengths are calculated.

5. The fifth reason reveals an even more Platonic view of existence than Plato himself expressed. Pacioli states that just as God conferred being to the entire cosmos through the fifth essence, represented by the dodecahedron, so does the Golden Ratio confer being to the dodecahedron, since one cannot construct the dodecahedron without the Golden Ratio. He adds that it is impossible to compare the other four Platonic solids (representing earth, water, air, and fire) to each other without the Golden Ratio.

In the book itself, Pacioli raves ceaselessly about the properties of the Golden Ratio. He analyzes in succession what he calls the thirteen different "effects" of the "divine proportion" and attaches to each one of these "effects" adjectives like "essential," "singular," "wonderful," "supreme," and so on. For example, he regards the "effect" that Golden Rectangles can be inscribed in the icosahedron (Figure 22) as "incomprehensible." Pacioli stops at thirteen "effects," concluding that, "for the sake of salvation, the list must end," because thirteen men were

present at the table at the Last
Supper.

There is no doubt that Pacioli had a great interest in
the arts and that his intention
in *The Divine Proportion* was
partly to perfect their mathematical basis. His opening
statement on the book's first
page expresses his desire to
reveal to artists, through the
Golden Ratio, the "secret" of
harmonic forms. To ensure its
attractiveness, Pacioli secured
for *The Divine Proportion* the
services of the dream illustrator of any author—Leonardo
da Vinci himself provided
sixty illustrations of solids,

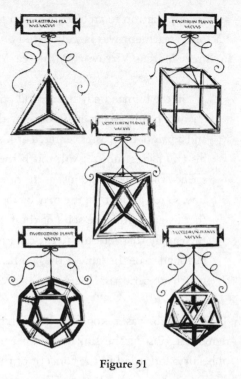

Figure 51

depicted in both skeletal (Figure 51) and solid forms (Figure 52).
Pacioli was quick to express his gratitude; he wrote about Leonardo's
contribution: "the most excellent painter in perspective, architect,
musician, the man endowed with all virtues, Leonardo da Vinci, who
deduced and elaborated a series of diagrams of regular solids." The
text itself, however, falls somewhat short of its declared high goals.

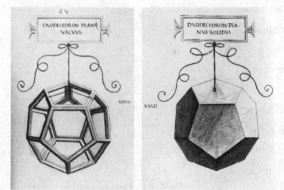

Figure 52

While the book starts
with a sensational flour-
ish, it continues with a
rather conventional set
of mathematical formu-
lae loosely wrapped up
in philosophical defini-
tions.

The second book in
the *Divina Proportione* is
a treatise on proportion

and its application to architecture and the structure of the human body. Pacioli's treatment was largely based on the work of the eclectic Roman architect Marcus Vitruvius Pollio (ca. 70–25 B.C.). Vitruvius wrote:

> . . . in the human body the central point is naturally the navel. For if a man be placed flat on his back, with his hands and feet extended, and a pair of compasses centered at his navel, the fingers and toes of his two hands and feet will touch the circumference of a circle described therefrom. And just as the human body yields a circular outline, so too a square figure may be found from it. For if we measure the distance from the soles of the feet to the top of the head, and then apply that measure to the outstretched arms, the breadth will be found to be the same as the height, as in the case of plane surfaces which are perfectly square.

This passage was taken by the Renaissance scholars as yet another demonstration of the link between the organic and geometrical basis of beauty, and it led to the concept of the "Vitruvian man," drawn beautifully by Leonardo (Figure 53; currently in the Galleria dell' Accademia, Venice). Accordingly, Pacioli's book also starts with a discussion of proportions in the human body, "since in the human body every sort of proportion and proportionality can be found, produced at the beck of the all-Highest through the inner mysteries of nature." However, contrary to frequent claims in the literature, Pacioli does not insist on the Golden Ratio as determining the proportions of all works of art. Rather, when dealing with design and proportion, he specifically advocates the Vitruvian

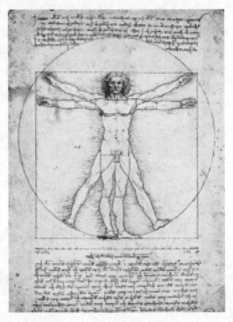

Figure 53

system, which is based on simple (rational) ratios. Author Roger Herz-Fischler traced the fallacy of the Golden Ratio as Pacioli's canon for proportion to a false statement made in the 1799 edition of *Histoire de Mathématiques* (History of mathematics) by the French mathematicians Jean Etienne Montucla and Jérôme de Lalande.

The third volume of the *Divina* (A short book divided into three partial tracts on the five regular bodies) is essentially an Italian word-by-word translation of Piero's *Five Regular Solids* composed in Latin. The fact that nowhere in the text does Pacioli acknowledge that he was merely the translator of the book provoked a violent denunciation from art historian Giorgio Vasari. Vasari writes about Piero della Francesca that he

> . . . was regarded as a great master of the problems of regular solids, both arithmetical and geometrical, but he was prevented by the blindness that overtook him in his old age, and then by death, from making known his brilliant researches and the many books he had written. The man who should have done his utmost to enhance Piero's reputation and fame, since Piero taught him all he knew, shamefully and wickedly tried to obliterate his teacher's name and to usurp for himself the honor which belonged entirely to Piero; for he published under his own name, which was Fra Luca dal Borgo [Pacioli], all the researches done by that admirable old man, who was a great painter as well as an expert in the sciences.

So, was Pacioli a plagiarist? Quite possibly, although in *Summa* he did render homage to Piero, whom he regarded as "the monarch of our times in painting" and one who "is familiar to you in that copious work which he composed on the art of painting and on the force of the line in perspective."

R. Emmett Taylor (1889–1956) published in 1942 a book entitled *No Royal Road: Luca Pacioli and His Times.* In this book, Taylor adopts a very sympathetic attitude toward Pacioli, and he argues that, on the basis of style, Pacioli may have had nothing to do with the third book of the *Divina* and it was just appended to Pacioli's work.

Be that as it may, there is no question that if not for Pacioli's *printed* books, Piero's ideas and mathematical constructions (which were not published in printed form) would not have reached the wide circulation that they eventually achieved. Furthermore, up until Pacioli's time, the Golden Ratio had been known only by rather intimidating names, such as "extreme and mean ratio" or "proportion having a mean and two extremes," and the concept itself was familiar only to mathematicians. The publication of *The Divine Proportion* in 1509 gave a new topical interest to the Golden Ratio. The concept could now be considered with fresh attention, because its publication in book form identified it as worthy of respect. The infusion of theological/philosophical meaning into the *name* ("Divine Proportion") also singled out the Golden Ratio as a mathematical topic into which an increasingly eclectic group of intellectuals could delve. Finally, with Pacioli's book, the Golden Ratio started to become available to artists in theoretical treatises that were not overly mathematical, that they could actually use.

Leonardo da Vinci's drawings of polyhedra for *The Divine Proportion,* drawn (in Pacioli's words) with his "ineffable left hand," had their own impact. These were probably the first illustrations of skeletal solids, which allowed for an easy visual distinction between front and back. Leonardo may have drawn the polyhedra from a series of wooden models, since records of the Council Hall in Florence indicate that a set of Pacioli's wooden models was purchased by the city for public display. In addition to the diagrams for Pacioli's book, we can find sketches of many solids scattered throughout Leonardo's notebooks. In one place he presents an approximate geometrical construction of the pentagon. This fusion of art and mathematics reaches its climax in Leonardo's *Trattato della pittura* (Treatise on painting; organized by Francesco Melzi, who inherited Leonardo's manuscripts), which opens with the admonition: "Let no one who is not a mathematician read my works"—hardly a likely statement to be found in any contemporary art handbook!

The drawings of solids in the *Divina* have also inspired some of the *intarsia* constructed by Fra Giovanni da Verona around 1520. Intarsia represent a special art form, in which elaborate flat mosaics are constructed of pieces of inlaid wood. Fra Giovanni's intarsia panels include an icosahedron, which almost certainly used Leonardo's skeletal drawing as a template.

The lives of Leonardo and Pacioli continued to be somewhat intertwined even after the completion of *The Divine Proportion*. In October of 1499 the two men fled Milan, after the French army, led by King Louis XII, captured that city. After spending brief periods of time in Mantua and Venice, both settled for some time in Florence. During the period of their friendship, Pacioli's name became associated with two other major mathematical works—a translation into Latin of Euclid's *Elements* and an unpublished work on recreational mathematics. Pacioli's translation of the *Elements* was an annotated version, based on an earlier translation by Campanus of Novara (1220–1296), which appeared in printed form in Venice in 1482 (and which was the first *printed* version). Pacioli did not manage to publish his compilation of problems in recreational mathematics and proverbs *De Viribus Quantitatis* (The powers of numbers) before his death in 1517. This work was a collaborative project between Pacioli and Leonardo, and Leonardo's own notes contain many of the problems in *De Viribus*.

Fra Luca Pacioli certainly cannot be remembered for originality, but his influence on the development of mathematics in general, and on the history of the Golden Ratio in particular, cannot be denied.

MELANCHOLY

Another major Renaissance figure who entertained an intriguing combination of artistic and mathematical interests is the German painter Albrecht Dürer.

Dürer is considered by many to be the greatest German artist of the Renaissance. He was born on May 21, 1471, in the Imperial Free City of Nürnberg, to a hardworking jeweler. At age nineteen, he already demonstrated talents and ability as a painter and woodcut designer that surpassed those of his teacher, the leading painter and book illustrator in Nürnberg, Michael Wolgemut. Dürer therefore embarked on four years of travel, during which he became convinced that mathematics, "the most precise, logical, and graphically constructive of the sciences," has to be an important ingredient of art.

Consequently, after a short stay in Nürnberg, during which he married Agnes Frey, the daughter of a successful craftsman, he left again for Italy, with the goal of expanding both his artistic and mathematical horizons. His visit to Venice in 1494–1495 seems to have accomplished precisely that. Dürer's meeting with the founder of the Venetian School of painting, Giovanni Bellini (ca. 1426–1516), left a great impression on the young artist, and his admiration for Bellini persisted throughout his life. At the same time, Dürer's encounter with Jacopo de' Barbari, who painted the wonderful portrait of Luca Pacioli (Figure 50), acquainted him with Pacioli's mathematical work and its relevance for art. In particular, de' Barbari showed Dürer two figures, male and female, that were constructed by geometrical methods, and the experience motivated Dürer to investigate human movement and proportions. Dürer probably met with Pacioli himself in Bologna, during a second visit to Italy in 1505 to 1507. In a letter from that period, he describes his visit to Bologna as being "for art's sake, for there is one there who will instruct me in the secret art of perspective." The mysterious "one" in Bologna has been interpreted by many as referring to Pacioli, although other names, such as those of the outstanding architect Donato di Angelo Bramante (1444–1514) and the architectural theorist Sebastiano Serlio (1475–1554), have also been suggested. During the same Italian trip Dürer also met again with Jacopo de' Barbari. This second visit, though, was marked by Dürer's somewhat paranoiac nervousness about harm that might be done to him by artists envious of his fame. For example, he refused invitations to dinner for fear that someone might try to poison him.

Starting in 1495, Dürer showed a serious interest in mathematics. He spent much time studying the *Elements* (a Latin translation of which he obtained in Venice, although he spoke little Latin), Pacioli's works on mathematics and art, and the important works on architecture, proportion, and perspective by the Roman architect Vitruvius and by the Italian architect and theorist Leon Baptista Alberti (1404–1472).

Dürer's contributions to the history of the Golden Ratio come both in the form of written work and through his art. His major treatise, *Unterweisung der Messung mit dem Zirkel und Richtscheit* (Treatise on measurement with compass and ruler), was published in 1525 and was one of the first books on mathematics published in German. In it Dürer complains that too many artists are ignorant of geometry, "without which no one can either be or become an absolute artist." The first of the four books of the *Treatise* gives detailed descriptions of the construction of various curves, including the logarithmic (or equiangular) spiral, which is, as we have seen, closely related to the Golden Ratio. The second book contains precise and approximate methods for the construction of many polygons, including two constructions of the pentagon (one exact and one approximate). The Platonic solids, as well as other solids, some of Dürer's own invention, together with the theory of perspective and of shadows, are discussed in the fourth book. Dürer's book was not intended to be used as a textbook of geometry—for example, he gives only one example of a proof. Rather, Dürer always starts with a practical application and then continues with an exposition of the very basic theoretical aspects. The book contains some of the earliest presentations of nets of polyhedra. These are plane sheets on which the surfaces of the polyhedra are drawn in such a way that the figures can be cut out (as single pieces) and folded to form the three-dimensional solids.

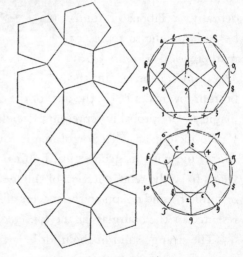

Dürer's illustration for the net of a dodecahedron (related as we know to the Golden Ratio) is shown in Figure 54.

Figure 54

Dürer mingled his virtuosity in woodcuts and engravings with his interest in mathematics in the enigmatic allegory "Melencolia I" (Figure 55). This is one of the trio of master engravings (the other two being "Knight, Death and Devil," and "St. Jerome in His Study"). It has been suggested that Dürer created the picture in a fit of melancholy after the death of his mother. The central figure in "Melencolia" is a winged female seated listless and dispirited on a stone ledge. In her right hand she holds a compass, opened for measuring. Most of the objects in the engraving have multiple symbolic meanings, and entire articles have been devoted to their interpretation. The pot on the fire in the middle left and the scale at the top are thought to represent alchemy. The "magic square" on the upper right (in which every row, column, diagonal, the four central numbers, and the numbers in the four corners add up to 34; incidentally, a Fibonacci number) is thought to represent mathematics (Figure 56). The middle entries in the

Figure 55

bottom row make 1514, the date of the engraving. The inclusion of the magic square probably represents Pacioli's influence, since Pacioli's *De Viribus* included a collection of magic squares. The main purport of the engraving, with its geometrical figures, keys, bat, seascape, and so on, seems to be the representation of the melancholy that engulfs the artist or thinker, amid doubts in the success of her endeavors, while time, represented by the hourglass at the top, goes on.

The strange solid in the middle left of the engraving has been the topic of serious discussion and various reconstruction attempts. At first sight it looks like a cube from which two opposite corners have been sliced off (which inspired some Freudian interpretations), but this appears not to be the case. Most researchers conclude that the figure is what is known as a rhombohedron (a six-sided solid with each side shaped as a rhombus; Figure 57), which has been truncated so that it can be circumscribed by a sphere. When resting on one of its triangular faces, its front fits precisely into the magic square. The angles in the face of the solid have also been a matter of some debate. While many sug-

Figure 56

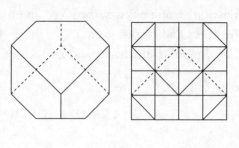

Figure 57

gest 72 degrees, which would relate the figure to the Golden Ratio (see Figure 25), Dutch crystallographer C. H. MacGillavry concluded on the basis of perspectival analysis that the angles are of 80 degrees. The perplexing qualities of this solid are summarized beautifully in an article by T. Lynch that appeared in 1982 in the *Journal of the Warburg and Courtauld Institutes*. The author concludes: "As a representation of polyhedra was seen as one of the main problems of perspective geometry, what better way could Dürer prove his ability in this field, than to include in an engraving a shape that was so new and perhaps even unique, and to leave the question of what it was, and where it came from, for other geometricians to solve?"

With the exception of the influential work of Pacioli and the mathematical/artistic interpretations of the painters Leonardo and Dürer, the sixteenth century brought about no other surprising developments in the story of the Golden Ratio. While a few mathematicians, including the Italian Rafael Bombelli (1526–1572) and the Spanish Franciscus Flussates Candalla (1502–1594) used the Golden Ratio in a variety of problems involving the pentagon and the Platonic solids, the more exciting applications had to await the very end of the century.

However, the works of Pacioli, Dürer, and others revived the interest in Platonism and Pythagoreanism. Suddenly the Renaissance intellectuals saw a real opportunity to relate mathematics and rational logic to the universe around them, in the spirit of the Platonic worldview. Concepts like the "Divine Proportion" built, on one hand, a bridge between mathematics and the workings of the cosmos and, on the other, a relation among physics, theology, and metaphysics. The person who, in his ideas and works, exemplifies more than any other this fascinating blending of mathematics and mysticism is Johannes Kepler.

MYSTERIUM COSMOGRAPHICUM

Johannes Kepler is best remembered as an outstanding astronomer responsible (among other things) for the three laws of planetary motion that bear his name. But Kepler was also a talented mathematician, a speculative metaphysician, and a prolific author. Born at a time of great political upheaval and religious chaos, Kepler's education, life, and thinking were critically shaped by the events around him. Kepler was born on December 27, 1571, in the Imperial Free City of Weil der Stadt, Germany, in his grandfather Sebald's house. His father, Heinrich, a mercenary soldier, was absent from home throughout most of Kepler's childhood, and during his short visits he was (in Kepler's words): "a wrongdoer, abrupt and quarrelsome." The father left home when Kepler was about sixteen, never to be seen again. He is supposed to have participated in a naval war for the Kingdom of Naples and to have died on his way home. Consequently, Kepler was raised mostly by his mother, Katharina, who worked in her father's inn. Katharina herself was a rather strange and unpleasant woman, who gathered herbs and believed in their magical healing powers. A series of events involving personal grudges, unfortunate gossip, and greed eventually led to her arrest at old age in 1620, and to an indictment of witchcraft. Such accusations were not uncommon at that time—no fewer than thirty-eight women were executed for witchcraft in Weil der Stadt in the years between 1615 and 1629. Kepler, who was already well known at the time of her arrest, reacted to the news of his mother's trial "with unutterable distress." He effectively took charge of her defense, enlisting the help of the legal faculty at the University of Tübingen. The charges against Katharina Kepler were eventually dismissed after a long ordeal, mainly in light of her own testimony under the threat of great pain and torture. This story conveys the atmosphere and the intellectual confusion that prevailed during the period of Kepler's scientific work. Kepler was born into a society that experienced (only fifty years earlier) Martin Luther's breaking with the Catholic church, proclaiming that humans' sole justification before God was faith. That society was also about to embark on the bloody and insane conflict known as the Thirty Years' War. We can only be astonished how, with this background and with the violent ups and downs of his tumultuous life, Kepler was able to produce a discovery that is regarded by many as the true birth of modern science.

Kepler started his studies at the higher seminary at Maulbronn and then won a scholarship from the Duke of Württemberg to attend the Lutheran seminary at the University of Tübingen in 1589. The two topics that attracted him most, and which in his mind were closely related, were theology and mathematics. At that time astronomy was considered a part of mathematics, and Kepler's teacher of astronomy was the prominent astronomer Michael Mästlin (1550–1631), with whom he continued to maintain contact even after leaving Tübingen.

In his formal lessons, Mästlin must have taught only the traditional Ptolemaic or geocentric system, in which the Moon, Mercury, Venus, the Sun, Mars, Jupiter, and Saturn all revolved around the stationary Earth. Mästlin, however, was fully aware of Nicolaus Copernicus' heliocentric system, which was published in 1543, and in private he did discuss the merits of such a system with his favorite student, Kepler. In the Copernican system, six planets (including Earth, but not including the Moon, which was no longer considered a planet but rather a "satellite") revolved around the Sun. In the same way that from a moving car you can observe only the relative motions of the other cars, in the Copernican system, much of what appears to be the motion of the planets simply reflects the motion of Earth itself.

Kepler seems to have taken an immediate liking to the Copernican system. The fundamental idea of this cosmology, that of a central Sun surrounded by a sphere of the fixed stars with a space between the sphere and the Sun, fit perfectly into his view of the cosmos. Being a profoundly religious person, Kepler believed that the universe represents a reflection of its Creator. The unity of the Sun, the stars, and the intervening space symbolized to him an equivalence to the Holy Trinity of the Father, Son, and the Holy Spirit.

While Kepler graduated with distinction from the faculty of arts and was close to finishing his theological studies, something happened to change his profession from that of a pastor to that of a mathematics teacher. The Protestant seminary in Graz, Austria, asked the University of Tübingen to recommend a replacement for one of their math teachers who had passed away, and the university selected Kepler. In March of 1594 Kepler therefore began, unwillingly, a month-long trip to Graz, in the Austrian province of Styria.

Realizing that fate had forced upon him the career of a mathematician, Kepler became determined to fulfill what he regarded as his Christian duty—to understand God's creation, the universe. Accordingly, he delved into the translations of the *Elements* and the works of the Alexandrian geometers Apollonius and Pappus. Accepting the general principle of the Copernican heliocentric system, he set out to search for answers to the following two major questions: Why were there precisely six planets? and What was it that determined that the planetary orbits would be spaced as they are? These "why" and "what" questions were entirely new in the astronomical vocabulary. Unlike the astronomers before him, who satisfied themselves with simply recording the observed positions of the planets, Kepler was seeking a theory that would explain it all. He expressed this new approach to human inquiry beautifully:

> In all acquisition of knowledge it happens that, starting out from those things which impinge on the senses, we are carried by the operation of the mind to higher things which cannot be grasped by any sharpness of the senses. The same thing happens also in the business of astronomy, in which we first of all perceive with our eyes the various positions of the planets at different times, and reasoning then imposes itself on these observations and leads the mind to recognition of the form of the universe.

But, wondered Kepler, what tool would God use to design His universe?

The first glimpse of what was to become his preposterously fantastic explanation to these cosmic questions dawned on Kepler on July

19, 1595, as he was trying to explain
the conjunctions of the outer planets,
Jupiter and Saturn (when the two bodies have the same celestial coordinate).
Basically, he realized that if he inscribed an equilateral triangle within
a circle (with its vertices lying on the
circle) and another circle inside the
triangle (touching the midpoints of
the sides; Figure 58), then the ratio of
the radius of the larger circle to that

Figure 58

of the smaller one was about the same as the ratio of the sizes of Saturn's
orbit to Jupiter's orbit. Continuing with this line of thought, he decided that to get to the orbit of Mars (the next planet closer to the Sun),
he would need to use the next geometrical figure—a square—inscribed
inside the small circle. Doing this, however, did not produce the right
size. Kepler did not give up, and being already along a path inspired by
the Platonic view, that "God ever geometrizes," it was only natural for
him to take the next geometrical step and try three-dimensional figures.
The latter exercise resulted in Kepler's first use of geometrical objects
related to the Golden Ratio.

Kepler gave the answer to the two questions that intrigued him in
his first treatise, known as *Mysterium Cosmographicum* (The cosmic mystery), which was published in 1597. The full title, given on the title
page of the book (Figure 59; although the publication date reads 1596,
the book was published the following year) reads: "A precursor to cos-
mographical dissertations, containing the cosmic mystery of the admirable proportions of the Celestial Spheres, and of the True and Proper
Causes of their Numbers, Sizes, and Periodic Motions of the Heavens,
Demonstrated by the Five Regular Geometric Solids."

Kepler's answer to the question of why there were six planets was
simple: because there are precisely five regular Platonic solids. Taken as
boundaries, the solids determine six spacings (with an outer spherical
boundary corresponding to the heaven of the fixed stars). Furthermore,
Kepler's model was designed so as to answer at the same time the question of the sizes of the orbits as well. In his words:

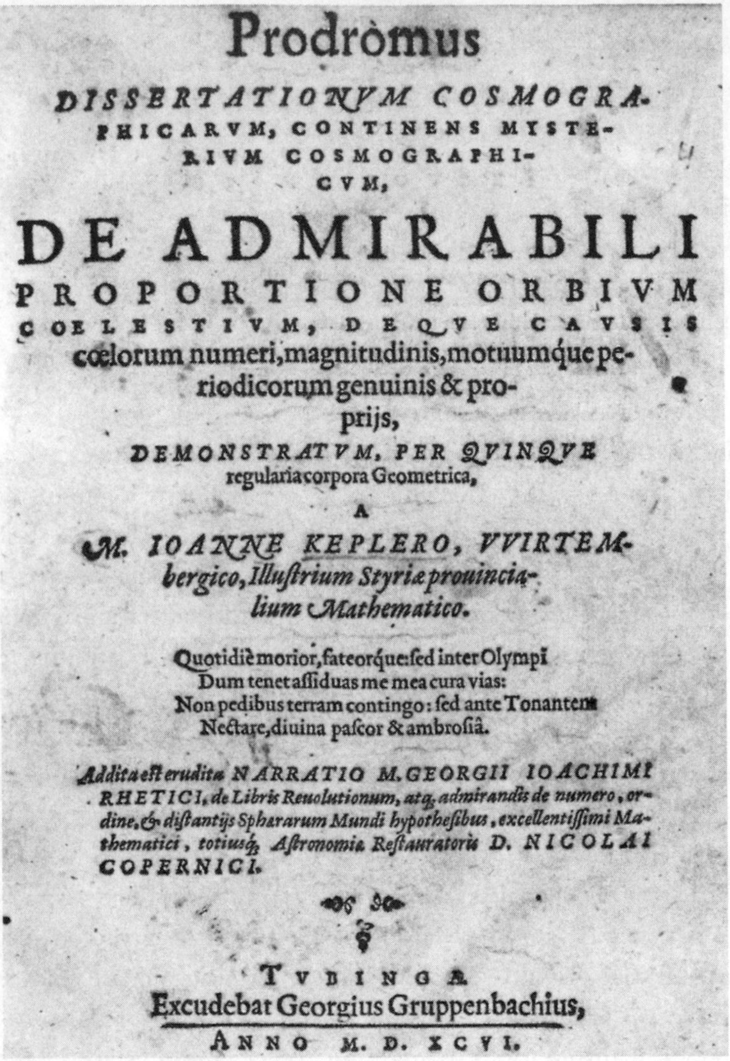

Prodròmus

DISSERTATIONVM COSMOGRA-
PHICARVM, CONTINENS MYSTE-
RIVM COSMOGRAPHI-
CVM,

DE ADMIRABILI

PROPORTIONE ORBIVM

COELESTIVM, DEQVE CAVSIS
cœlorum numeri, magnitudinis, motuumque pe-
riodicorum genuinis & pro-
prijs,

DEMONSTRATVM, PER QVINQVE
regularia corpora Geometrica,

A

M. IOANNE KEPLERO, VVIRTEM-
bergico, Illustrium Styriæ prouincia-
lium Mathematico.

Quotidiè morior, fateorque: sed inter Olympi
Dum tenet assiduas me mea cura vias:
Non pedibus terram contingo: sed ante Tonantem
Nectare, diuina pascor & ambrosiâ.

Addita est erudita NARRATIO M. GEORGII IOACHIMI
RHETICI, de Libris Reuolutionum, atq, admirandis de numero, or-
dine, & distantijs Sphararum Mundi hypothesibus, excellentissimi Ma-
thematici, totiusq, Astronomia Restauratoris D. NICOLAI
COPERNICI.

TVBINGÆ
Excudebat Georgius Gruppenbachius,
ANNO M. D. XCVI.

Figure 59

The Earth's sphere is the measure of all other orbits. Circumscribe a dodecahedron around it. The sphere surrounding it will be that of Mars. Circumscribe a tetrahedron around Mars. The sphere surrounding it will be that of Jupiter. Circumscribe a cube around Jupiter. The surrounding sphere will be that of Saturn. Now, inscribe an icosahedron inside the orbit of the Earth. The sphere inscribed in it will be that of Venus. Inscribe an octahedron inside

Venus. The sphere inscribed in it will be that of Mercury. There you
have the basis for the number of the planets.

Figure 60 shows a schematic from *Mysterium Cosmographicum,* which illustrates Kepler's cosmological model. Kepler explained at some length
why he made the particular associations between the Platonic solids and
the planets, on the basis of their geometrical, astrological, and metaphysical attributes. He ordered the solids based on relationships to the
sphere, assuming that the differences between the sphere and the other
solids reflected the distinction between the creator and his creations.
Similarly, the cube is characterized by a *single* angle—the right angle.
To Kepler this symbolized the solitude associated with Saturn, and so
on. More generally, astrology was relevant to Kepler because "man is
the goal of the universe and of all creation," and the
metaphysical approach was justified by the fact that
"the mathematical things are the causes of the physical because God from the beginning of time carried
within himself in simple and divine abstraction the
mathematical objects as proto-
types for the materially planned
quantities."

Earth's position was chosen
so as to separate the solids that
can stand upright (i.e., cube,
tetrahedron, and dodecahedron),
from those that "float" (i.e., octa-
hedron and icosahedron).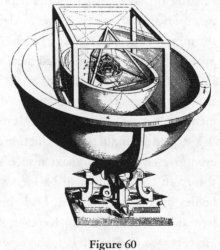

The spacings of the planets
resulting from this model agreed
reasonably well for some planets
but were significantly discrepant
for others (although the discrep-   Figure 60
ancies were usually no more than 10 percent). Kepler, absolutely
convinced of the correctness of his model, attributed most of the inconsistencies to inaccuracies in the measured orbits. He sent copies of the
book to various astronomers for comments, including a copy to one of

the foremost figures of the time, the Danish Tycho Brahe (1546–1601). One copy even made it into the hands of the great Galileo Galilei (1564–1642), who informed Kepler that he too believed in Copernicus' model but lamented the fact that "among a vast number (for such is the number of fools)" Copernicus "appeared fit to be ridiculed and hissed off the stage."

Needless to say, Kepler's cosmological model, which was based on the Platonic solids, was not only absolutely wrong, but it was crazy even for Kepler's time. The discovery of the planets Uranus (next after Saturn in terms of increasing distance from the Sun) in 1781 and Neptune (next after Uranus) in 1846 put the final nails into the coffin of an already moribund idea. Nevertheless, the importance of this model in the history of science cannot be overemphasized. As astronomer Owen Gingerich has put it in his biographical article on Kepler: "Seldom in history has so wrong a book been so seminal in directing the future course of science." Kepler took the Pythagorean idea of a cosmos that can be explained by mathematics a huge step forward. He developed an actual *mathematical model* for the universe, which on one hand was based on existing observational measurements and on the other was *falsifiable* by observations that could be made subsequently. These are precisely the ingredients required by the "scientific method"—the organized approach to explaining observed facts with a model of nature. An idealized scientific method begins with the collection of facts, a model is then proposed, and the model's predictions are tested through experiments or further observations. This process is sometimes summed up by the sequence: induction, deduction, verification. In fact, Kepler was even given a chance to make a successful prediction on the basis of his theory. In 1610, Galileo discovered with his telescope four new celestial bodies in the Solar System. Had these proven to be planets, it would have dealt a fatal blow to Kepler's theory already during his lifetime. However, to Kepler's relief, the new bodies turned out to be satellites (like our Moon) around Jupiter, not new planets revolving around the Sun.

Present-day physical theories that aim at explaining the existence of all the elementary (subatomic) particles and the basic interactions among them rely on mathematical symmetries in a very similar fashion to Kepler's theory relying on the symmetry properties of the Platonic solids to explain the number and properties of the planets. Kepler's model had something else in common with today's fundamental theory of the universe: Both theories are by their very nature *reductionistic*—they attempt to explain many phenomena in terms of a few fundamental laws. For example, Kepler's model deduced both the number of planets and the properties of their orbits from the Platonic solids. Similarly, modern theories known as string theories use basic entities (strings) which are extremely tiny (more than a billion billion times smaller than the atomic nucleus) to deduce the properties of all the elementary particles. Like a violin string, the strings can vibrate and produce a variety of "tones," and all the known elementary particles simply represent these different tones.

Kepler's continued interest in the Golden Ratio during his stay in Graz produced another interesting result. In October 1597, he wrote to Mästlin, his former professor, about the following theorem: "If on a line which is divided in extreme and mean ratio one constructs a right angled triangle, such that the right angle is on the perpendicular put at the section point, then the smaller leg will equal the larger segment of the divided line." Kepler's statement is represented by Figure 61. Line *AB* is divided in a Golden Ratio by point *C*. Kepler constructs a right-angled triangle *ADB* on *AB* as a hypotenuse, with the right angle *D* being on the perpendicular put at the Golden Section point *C*. He then proves that *BD* (the shorter side of the right angle) is equal to *AC* (the longer segment of the line divided in Golden Ratio). What makes this particular triangle special (other than the use of the Golden Ratio) is that in 1855 it was used by pyramidologist Friedrich Röber in one of

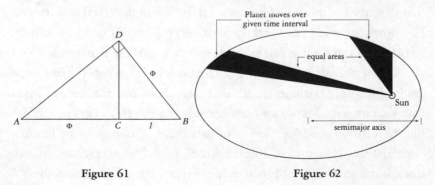

Figure 61   Figure 62

the false theories explaining the appearance of the Golden Ratio in the design of the pyramids. Röber was not aware of Kepler's work, but he used a similar construction to support his view that the "divine proportion" played a crucial role in architecture.

Kepler's *Mysterium Cosmographicum* led to a meeting between him and Tycho Brahe in Prague—at the time the seat of the Holy Roman Emperor. The meeting took place on February 4, 1600, and was the prelude to Kepler's moving to Prague as Tycho's assistant in October of the same year (after being forced out of Catholic Graz because of his Lutheran faith). When Brahe died on October 24, 1601, Kepler became the Imperial Mathematician.

Tycho left a huge body of observations, in particular of the orbit of Mars, and Kepler used these data to discover the first two laws of planetary motions named after him. Kepler's First Law states that the orbits of the known planets around the Sun are not exact circles but rather ellipses, with the Sun at one focus (Figure 62; the elongation of the ellipse is greatly exaggerated). An ellipse has two points called foci, such that the sum of the distances of any point on the ellipse from the two foci is the same. Kepler's Second Law establishes that the planet moves fastest when it is closest to the Sun (the point known as perihelion) and slowest when it is farthest (aphelion), in such a way that the line joining the planet to the Sun sweeps equal areas in equal time intervals (Figure 62). The question of what causes Kepler's laws to hold true was the outstanding unsolved problem of science for almost seventy years after Kepler published the laws. It took the genius of Isaac Newton (1642–1727) to deduce that the force holding the planets in their orbits is gravity. Newton explained Kepler's laws by solving together the laws that describe the motion of bodies with the law of universal gravitation. He showed that elliptical orbits with varying speeds (as described by Kepler's laws) represent one possible solution to these equations.

Kepler's heroic efforts in the calculations of Mars' orbit (many hundreds of sheets of arithmetic and their interpretation; dubbed by him as "my warfare with Mars") are considered by many researchers as signifying the birth of modern science. In particular, at one point he found a circular orbit that matched nearly all of Tycho's observations. In two cases, however, this orbit predicted a position that differed from the observations by about a quarter of the angular diameter of a full moon. Kepler wrote about this event: "If I had believed that we could ignore these eight minutes [of arc], I would have patched up my hypothesis in Chapter 16 accordingly. Now, since it was not permissible to disregard, those eight minutes alone pointed the path to a complete reformation in astronomy."

Kepler's years in Prague were extremely productive in both astronomy and mathematics. In 1604, he discovered a "new" star, now known as Kepler's Supernova. A supernova is a powerful stellar explosion, in which a star nearing the end of its life ejects its outer layers at a speed of ten thousand miles per second. In our own Milky Way galaxy, one such explosion is expected to occur on the average every one hundred years. Indeed, Tycho discovered a supernova in 1572 (Tycho's Supernova), and Kepler discovered one in 1604. Since then, however, for unclear reasons, no other supernova has been discovered in the Milky Way (although one exploded apparently unnoticed in the 1660s). Astronomers remark jokingly that maybe this paucity of supernovae simply reflects the fact that there have been no truly great astronomers since Tycho and Kepler.

In June 2001, I visited the house in which Kepler lived in Prague, at 4 Karlova Street. Today, this is a busy shopping street, and it is easy to miss the rusty plaque above the number 4, which states that Kepler lived there from 1605 to 1612. One of the shop owners just below Kepler's apartment did not even know that one of the greatest astronomers of all times had lived there. The rather sad-looking inner courtyard does contain a small sculpture of the armillary sphere with Kepler's name written across it, and another plaque is located near the mailboxes. Kepler's apartment itself, however, is not marked in any special way and is not open to the public, being occupied by one of the many families who live in the residential upper floors.

Kepler's mathematical work produced a few more highlights in the history of the Golden Ratio. In the text of a letter that he wrote in 1608 to a professor in Leipzig, we find that he discovered the relation between Fibonacci numbers and the Golden Ratio. He repeats the contents of that discovery in an essay tracing the reason for the six-cornered shape of snowflakes. Kepler writes:

Of the two regular solids, the dodecahedron and the icosahedron . . . both of these solids, and indeed the structure of the pentagon itself, cannot be formed without the divine proportion as the geometers of today call it. It is so arranged that the two lesser terms of a progressive series together constitute the third, and the two last, when added, make the immediately subsequent term and so on to infinity, as the same proportion continues unbroken . . . the further we advance from the number one, the more perfect the example becomes. Let the smallest numbers be 1 and 1 . . . add them, and the sum will be 2; add to this the latter of the 1's, result 3; add 2 to this, and get 5; add 3, get 8; 5 to 8, 13; 8 to 13, 21. As 5 is to 8, so 8 is to 13, approximately, and as 8 to 13, so 13 is to 21, approximately.

In other words, Kepler discovered that the ratio of consecutive Fibonacci numbers converges to the Golden Ratio. In fact, he also discovered another interesting property of the Fibonacci numbers: that the square of any term differs by 1 at most from the product of the two adjacent terms in the sequence. For example, since the sequence is: 1, 1, 2, 3, 5, 8, 13, 21, 34, . . . , if we look at $3^2 = 9$, it is only different by 1 from the product of the two terms that are adjacent to 3, $2 \times 5 = 10$. Similarly, $13^2 = 169$ is different by 1 from $8 \times 21 = 168$, and so on.

This particular property of Fibonacci numbers gives rise to a puzzling paradox first presented by the great creator of mathematical puzzles, Sam Loyd (1841–1911).

Consider the square of eight units on the side (area of $8^2 = 64$) in Figure 63. Now dissect it into four parts as indicated. The four pieces can be reassembled (Figure 64) to form a rectangle of sides 13 and 5

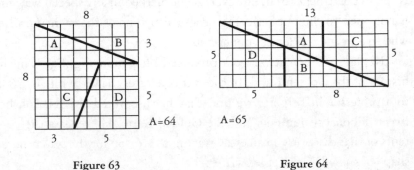

Figure 63   Figure 64

with an area of 65! Where did the extra square unit come from? The solution to the paradox is in the fact that the pieces actually do not fit exactly along the rectangle's long diagonal—there is a narrow space (a long thin parallelogram hidden under the thick line marking the long diagonal in Figure 64) with an area of one square unit. Of course, 8 is a Fibonacci number, and its square ($8^2 = 64$) differs by 1 from the product of its two adjacent Fibonacci numbers ($13 \times 5 = 65$)—the property discovered by Kepler.

You have probably noticed that Kepler refers to the Golden Ratio as "the divine proportion as the geometers of today call it." The combination of rational elements with Christian beliefs characterizes all of Kepler's endeavors. As a Christian natural philosopher, Kepler regarded it as his duty to understand the universe together with the intentions of its creator. Fusing his ideas on the Solar System with a strong affinity to the number 5, which he adopted from the Pythagoreans, Kepler writes about the Golden Ratio:

A peculiarity of this proportion lies in the fact that a similar proportion can be constructed out of the larger part and the whole; what was formerly the larger part now becomes the smaller, what was formerly the whole now becomes the larger part, and the sum of these two now has the ratio of the whole. This goes on indefinitely; the divine proportion always remaining. I believe that this geometrical proportion served as idea to the Creator when He introduced the creation of likeness out of likeness, which also continues indefinitely. I see the number five in almost all blossoms which lead the way for a fruit, that is, for creation, and which exist, not for their own sake, but for that of the fruit to follow. Almost all tree-blossoms can be included here; I must perhaps exclude lemons and oranges; although I have not seen their blossoms and am judging from the fruit or berry only which are not divided into five, but rather into seven, eleven, or nine cores. But in geometry, the number five, that is the pentagon, is constructed by means of the divine proportion which I wish [to assume to be] the prototype for the creation. Furthermore, there exists between the movement of the Sun (or, as I believe, the Earth) and that of Venus, which stands at the top of generative capability the ratio of 8 to 13 which, as we shall hear, comes very close to the divine proportion. Lastly, according to Copernicus, the Earth-sphere is midway between the spheres Mars and Venus. One obtains the proportion between them from the do-decahedron and the icosahedron, which in geometry are both deriv-atives of the divine proportion; it is on our Earth, however, that the act of procreation takes place.

Now see how the image of man and woman stems from the di-vine proportion. In my opinion, the propagation of plants and the progenitive acts of animals are in the same ratio as the geometrical proportion, or proportion represented by line segments, and the arithmetic or numerically expressed proportion.

Simply put, Kepler truly believed that the Golden Ratio served as a fundamental tool for God in creating the universe. The text also shows that Kepler was aware of the appearance of the Golden Ratio and Fi-bonacci numbers in the petal arrangements of flowers.

Kepler's relatively tranquil and professionally fruitful years in Prague ended in 1611 with a series of disasters. First, his son Friedrich died of smallpox, then his wife, Barbara, died of a contagious fever brought along by the occupying Austrian troops. Finally, Emperor Rudolph was deposed, abdicating the crown in favor of his brother Matthias, who was not known for his tolerance of Protestants. Kepler was therefore forced to leave for Linz in present-day Austria.

The crowning jewel of Kepler's work at Linz came in 1619, with the publication of his second major work on cosmology, *Harmonice Mundi* (Harmony of the world).

Recall that music and harmony represented to Pythagoras and the Pythagoreans the first evidence that cosmic phenomena could be de-scribed by mathematics. Only strings plucked at lengths with ratios corresponding to simple numbers produced consonant tones. A ratio of 2:3 sounded the fifth, 3:4 a fourth, and so on. Similar harmonic spac-ings of the planets were also thought to produce the "music of the spheres." Kepler was very familiar with these concepts since he read most of the book by Galileo's father, Vincenzo Galilei, *Dialogue Concern-ing Ancient and Modern Music,* although he rejected some of Vincenzo's ideas. Since he also believed that he had a complete model for the Solar System, Kepler was able to develop little "tunes" for the different planets (Figure 65).

Figure 65

As Kepler was convinced that "before the origin of things, geometry was coeternal with the Divine Mind," much of the *Harmony of the World* is devoted to geometry. One aspect of this work that is particularly important for the story of the Golden Ratio is Kepler's work on tiling, or tessellation.

In general, the word "tiling" is used to describe a pattern or structure that comprises of one or more shapes of "tiles" that pave a plane exactly, with no spaces, such as the arrangements in mosaics or floor tiles. In Chapter 8 we shall see that some of the mathematical concepts present in tiling are intimately related to the Golden Ratio. While Kepler was not aware of all the intricacies of the mathematics of tiling, his interest in the relationship between different geometrical forms and his admiration for the pentagon—the most direct manifestation of the "divine proportion"—was sufficient to lead him to interesting work on tiling. He was particularly interested in the congruence (fitting together) of geometrical shapes like polygons and solids. Figure 66 shows

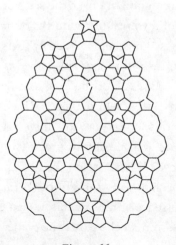

Figure 66

an example from *The Harmony of the World.* This particular tiling pattern is composed of four shapes, all related to the Golden Ratio: pentagons, pentagrams, decagons, and double decagons. To Kepler, this is a manifestation of "harmony," since *harmonia* in Greek means "a fitting together."

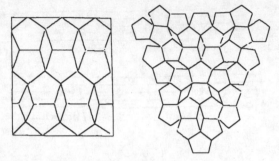

Figure 67

Interestingly, two other men who played significant roles in the history of the Golden Ratio before Kepler (and whose work was described in previous chapters) also showed interest in tiling—the tenth-century mathematician Abu'l-Wafa and the painter Albrecht Dürer. Both of them presented designs containing figures with fivefold symmetry. (An example of Dürer's work is shown in Figure 67.)

The fifth book of *Harmony of the World* contains Kepler's most significant result in astronomy—Kepler's Third Law of planetary motion. This represents the culmination of all of his agonizing over the sizes of the orbits of the planets and their periods of revolution around the Sun. Twenty-five years of work have been condensed into one incredibly simple law: The ratio of the period squared to the semimajor axis cubed is the same for all the planets (the semimajor axis is half the long axis of the ellipse; Figure 62). Kepler discovered this seminal law, which served as the basis for Newton's formulation of the law of universal gravitation, only when *Harmony of the World* was already in press. Unable to control his exhilaration he announced: "I have stolen the golden vessels of the Egyptians to build a tabernacle for my God from them, far away from the borders of Egypt." The essence of the law follows naturally from the law of gravity: The force is stronger the closer the planet is to the Sun, so inner planets must move faster to avoid falling toward the Sun.

In 1626, Kepler moved to Ulm and completed the *Rudolphine Tables,* the most extensive and accurate astronomical tables produced until that time. While I was visiting the University of Vienna in June 2001, my hosts showed me in the observatory's library a first edition of the tables (147 copies are known to exist today). The frontispiece of the book (Figure 68), a symbolic representation of the history of astronomy, contains at the lower left corner what may be Kepler's only self-portrait (Figure 69). It shows Kepler working by candlelight, under a banner listing his important publications.

Kepler died at noon on November 15, 1630, and was buried in Regensburg. Befitting his turbulent life, wars have totally destroyed his tomb, without a trace. Luckily, a sketch of the gravestone made by a friend survived, and it contains Kepler's epitaph:

Figure 68

Figure 69

I used to measure the heavens,
Now the Earth's shadows I measure
My mind was in the heavens,
Now the shadow of my body rests here.

Today, Kepler's originality and productivity are almost incomprehensible. We should realize that this was a man who endured unimaginable personal hardships, including the loss of three of his children in less than six months during 1617 and 1618. The English poet John Donne (1572–1631) perhaps described him best when he said that Kepler "hath received it into his care, that no new thing should be done in heaven without his knowledge."

7

PAINTERS AND POETS
HAVE EQUAL LICENSE

Painting isn't an aesthetic operation; it's a form of magic designed
as a mediator between this strange hostile world and us.
—PABLO PICASSO (1881–1973)

The Renaissance produced a significant change in direction in the history of the Golden Ratio. No longer was this concept confined to mathematics. Now the Golden Ratio found its way into explanations of natural phenomena and into the arts.

We have already encountered claims that the architectural design of various structures from antiquity, such as the Great Pyramid and the Parthenon, had been based on the Golden Ratio. A closer examination of these claims revealed, however, that in most cases they could not be substantiated. The introduction of the notion of the existence of a "Divine Proportion" and the general recognition of the importance of mathematics for perspective made it more conceivable that some artists would start using scientifically based methods in general and the Golden Ratio in particular in their works. Contemporary painter and draftsman David Hockney argues in his book *Secret Knowledge* (2001), for example, that starting with around 1430, artists began secretly using cameralike devices, including lenses, concave mirrors, and the camera obscura, to help them create realistic-looking paintings. But did artists really use the Golden Ratio? And if they did, was the Golden Ratio's application restricted to the visual arts or did it penetrate into other areas of artistic endeavor?

THE ARTIST'S SECRET GEOMETRY?

Many of the assertions concerning the employment of the Golden Ratio in painting are directly associated with the presumed aesthetic properties of the Golden Rectangle. I shall discuss the reality (or falsehood) of such a canon for aesthetics later in the chapter. For the moment, however, I shall concentrate on the much simpler question: Did any pre- and Renaissance painters actually base their artistic composition on the Golden Rectangle? Our attempt to answer this question takes us back to the thirteenth century.

The "Ognissanti Madonna" (also known as "Madonna in Glory," Figure 70; currently in the Uffizi Gallery in Florence) is one of the

Figure 70   Figure 71

greatest panel paintings by the famous Italian painter and architect Giotto di Bondone (1267–1337). Executed between 1306 and 1310, the painting shows a half-smiling, enthroned Virgin caressing the knee of the Child. The Madonna and Child are surrounded by angels and saints arranged in some sort of perspectival "hierarchy." Many books and articles on the Golden Ratio repeat the statement that both the painting as a whole and the central figures can be inscribed precisely in Golden Rectangles (Figure 71).

A similar claim is made about two other paintings with the same general subject: the "Madonna Rucellai" (painted in 1285) by the great Sienese painter Duccio di Buoninsegna, known as Duccio (ca. 1255–1319), and the "Santa Trinita Madonna" by the Florentine painter Cenni di Pepo, known as Cimabue (ca. 1240–1302). As fate would have it, currently the three paintings happen to be hanging in the same room in the Uffizi Gallery in Florence. The dimensions of the "Ognissanti," "Rucellai," and "Santa Trinita" Madonnas give height to width ratios of 1.59, 1.55, and 1.73, respectively. While all three numbers are not too far from the Golden Ratio, two of them are actually closer to the simple ratio of 1.6 rather than to the irrational number ϕ. This fact could indicate (if anything) that the artists followed the Vitruvian suggestion for a simple proportion, one that is the ratio of two whole numbers, rather than the Golden Ratio. The inner rectangle in the "Ognissanti Madonna" (Figure 71) leaves us with an equally ambiguous impression. Not only are the boundaries of the rectangle drawn usually (e.g., in Trudi Hammel Garland's charming book *Fascinating Fibonaccis*) with rather thick lines, making any measurement rather uncertain, but, in fact, the upper horizontal side is placed somewhat arbitrarily.

Remembering the dangers of having to rely on measured dimensions alone, we may wonder if there exist any other reasons to suspect that these three artists might have desired to include the Golden Ratio in their paintings. The answer to this question appears to be negative, unless they were driven toward this ratio by some unconscious aesthetic preference (a possibility that will be discussed later in the chapter). Recall that the three Madonnas were painted more than two centuries before the publication of *The Divine Proportion* brought the ratio to wider attention.

The French painter and author Charles Bouleau expresses a different view in his 1963 book *The Painter's Secret Geometry*. Without referring to Giotto, Duccio, or Cimabue specifically, Bouleau argues that Pacioli's book represented an end to an era rather than its beginning. He asserts that *The Divine Proportion* merely "reveals the thinking of long centuries of oral tradition" during which the Golden Ratio "was considered as the expression of perfect beauty." If this were truly the case, then Cimabue, Duccio, and Giotto indeed might have decided to use this accepted standard for perfection. Unfortunately, I find no evidence to support Bouleau's statement. Quite to the contrary; the documented history of the Golden Ratio is inconsistent with the idea that this proportion was particularly revered by artists in the centuries preceding the publication date of Pacioli's book. Furthermore, all the serious studies of the works of the three artists by art experts (e.g., *Giotto* by Francesca Flores D'Arcais; *Cimabue* by Luciano Bellosi) give absolutely no indication whatsoever that these painters might have used the Golden Ratio—the latter claim appears only in the writings of Golden Number enthusiasts and is based solely on the dubious evidence of measured dimensions.

Another name that invariably turns up in almost every claim of the appearance of the Golden Ratio in art is that of Leonardo da Vinci. Some authors even attribute the invention of the name "the Divine Proportion" to Leonardo. The discussion usually concentrates on five works by the Italian master: the unfinished canvas of "St. Jerome," the two versions of "Madonna of the Rocks," the drawing of "a head of an old man," and the famous "Mona Lisa." I am going to ignore the "Mona Lisa" here for two reasons: It has been the subject of so many volumes of contradicting scholarly and popular speculations that it would be virtually impossible to reach any unambiguous conclusions; and the Golden Ratio is supposed to be found in the dimensions of a rectangle around Mona Lisa's face. In the absence of any clear (and documented) indication of where precisely such a rectangle should be drawn, this idea represents just another opportunity for number juggling. I shall return, however, to the more general topic of proportions in faces in Leonardo's paintings, when I shall discuss the drawing "a head of an old man."

Figure 72   Figure 73

The case of the two versions of "Madonna of the Rocks" (one in the Louvre in Paris, Figure 72, and the other in the National Gallery in London, Figure 73) is not particularly convincing. The ratio of the height to width of the painting thought to have been executed earlier (Figure 72) is about 1.64 and of the later one 1.58, both reasonably close to φ but also close to the simple ratio of 1.6.

The dating and authenticity of the two "Madonna of the Rocks" also put an interesting twist on the claims about the presence of the Golden Ratio. Experts who studied the two paintings concluded that, without a doubt, the Louvre version was done entirely by Leonardo's hand, while the execution of the National Gallery version might have been a collaborative effort and is still the source of some debate. The Louvre version is thought to be one of the first works that Leonardo produced in Milan, probably between 1483 and 1486. The National Gallery painting, on the other hand, usually is assumed to have been completed around 1506. The reason that these dates may be of some significance is that Leonardo met Pacioli for the first time in 1496, in the Court of Milan. The seventy-first chapter of the *Divina* (the end of the first portion of the book) was, in Pacioli's words: "Finished this day of December 14, at Milan in our still cloister the year 1497." The first version (and the one with no doubts about authenticity) of the "Madonna of the Rocks" was therefore completed about ten years before Leonardo had the opportunity to hear directly from the horse's mouth about the "divine proportion." The claim that Leonardo used the Golden Ratio in "Madonna of the Rocks" therefore amounts to believing that the artist adopted this proportion even before he started his collaboration with Pacioli. While this is not impossible, there is no evidence to support such an interpretation.

Either version of "Madonna of the Rocks" represents one of Leonardo's most accomplished masterpieces. Perhaps in no other painting did he apply better his poetic formula: "every opaque body is surrounded and clothed on its surface by shadows and light." The figures in the paintings literally open themselves to the emotional participation of the spectator. To claim that these paintings derive any part of their strength from the mere ratio of their dimensions trivializes Leonardo's genius unnecessarily. Let us not fool ourselves; the feeling of awe we experience when facing "Madonna of the Rocks" has very little to do with whether the dimensions of the paintings are in a Golden Ratio.

A similar uncertainty exists with respect to the unfinished "St. Jerome" (Figure 74; currently in the Vatican museum). Not only is the painting dated to 1483, long before Pacioli's move to Milan, but the claim made in some books (e.g., in David Bergamini and the editors of *Life Magazine*'s *Mathematics*) that "a Golden Rectangle fits so neatly around St. Jerome" requires quite a bit of wishful thinking. In fact, the sides of the rectangle miss the body (especially on the left side) and head entirely, while the arm extends well beyond the rectangle's side.

The last example for a possible use of the Golden Ratio by Leonardo is the drawing of "a head of an old man" (Figure 75; the drawing is currently in the Galleria dell'Accademia in Venice). The profile and diagram of proportions were drawn in pen some time around 1490. Two studies of horsemen in red chalk, which are associated with Leonardo's "Battle of Anghiari," were added to the same page around 1503–1504.

While the overlying grid leaves very little doubt that Leonardo was indeed interested in various proportions in the face, it is very difficult to draw any definitive conclusions from this study. The rectangle in the middle left, for example, is approximately a Golden Rectangle, but the lines are drawn so roughly that we cannot be sure. Nevertheless, this drawing probably comes the closest to a demonstration that Leonardo used rectangles to determine dimensions in his paintings and that he might have even considered the application of the Golden Ratio to his art.

Leonardo's interest in proportions in the face may have another in-

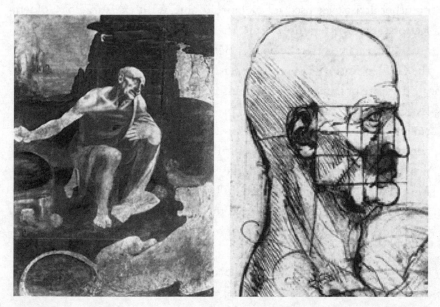

Figure 74   Figure 75

teresting manifestation. In an article that appeared in 1995 in the *Scientific American,* art historian and computer graphics artist Lillian Schwarz presented an interesting speculation. Schwarz claimed that in the absence of his model for the "Mona Lisa," Leonardo used his own facial features to complete the painting. Schwarz's suggestion was based on a computer-aided comparison between various dimensions in Mona Lisa's face and the respective dimensions in a red chalk drawing that is considered by many (but not all) to be Leonardo's only self-portrait.

However, as other art analysts have pointed out, the similarity in the proportions may simply reflect the fact that Leonardo used the same formulae of proportion (which may or may not have included the Golden Ratio) in the two portraits. In fact, Schwarz herself notes that even in his grotesques—a collection of bizarre faces with highly exaggerated chins, noses, mouths, and foreheads—Leonardo used the same proportions in the face as in the "head of an old man."

If there exist serious doubts regarding whether Leonardo himself, who was not only a personal friend of Pacioli but also the illustrator for the *Divina,* used the Golden Ratio in his paintings, does this mean that no other artist ever used it? Definitely not. With the surge of Golden Ratio academic literature toward the end of the nineteenth century, the artists also started to take notice. Before we discuss artists who did use the Golden Ratio, however, another myth still needs to be dispelled.

In spite of many existing claims to the contrary, the French pointillist Georges Seurat (1859–1891) probably did not use the Golden Ratio in his paintings. Seurat was interested in color vision and color combination, and he used the pointillist (multidotted) technique to approximate as best as he could the scintillating, vibratory quality of light. He was also concerned late in life with the problem of expressing specific emotions through pictorial means. In a letter he wrote in 1890, Seurat describes succinctly some of his views:

> Art is harmony. Harmony is the analogy of contradictions and of similars, in tone, shade, line, judged by the dominant of those and under the influence of a play of light in arrangements that are gay, light, sad. Contradictions are . . . , with respect to line, those that form a right angle. . . . Gay lines are lines above the horizontal; . . . calm is the horizontal; sadness lines in the downward direction.

Seurat used these ideas explicitly in "The Parade of a Circus" (sometimes called "The Side Show"; Figure 76; currently in The Metropolitan Museum of Art, New York). Note in particular the right angle formed by the balustrade and the vertical line to the right of the middle of the painting. The entire composition is based on principles that Seurat adopted from art theorist David Sutter's book *La philosophie des*

Beaux-Arts appliquée à la peinture (The philosophy of the fine arts applied to painting; 1870). Sutter wrote: "when the dominant is horizontal, a succession of vertical objects can be placed on it because this series will concur with the horizontal line."

Figure 76

Golden Ratio aficionados often present analyses of "The Parade" (as well as other paintings, such as "The Circus") to "prove" the use of φ. Even in the beautiful book *Mathematics,* by Bergamini and the editors of *Life Magazine,* we find: *"La Parade,* painted in the characteristic multi-dotted style of the French impressionist Georges Seurat, contains numerous examples of Golden proportions." The book goes even further with a quote (attributed to "one art expert") that Seurat "attacked every canvas by the Golden Section." Unfortunately, these statements are unfounded. This myth was propagated by the Romanian born prelate and author Matila Ghyka (1881–1965), who was also the "art expert" quoted by Bergamini. Ghyka published two influential books, *Esthétique des proportions dans la nature et dans les arts* (Aesthetics of proportions in nature and in the arts; 1927) and *Le Nombre d'Or: Rites et rythmes pytagoriciens dans le développement de la civilisation occidentale* (The golden number, Pythagorean rites and rhythms in the development of Western civilization; 1931). Both books are composed of semimystical interpretations of mathematics. Alongside correct descriptions of the mathematical properties of the Golden Ratio, the books contain a collection of inaccurate anecdotal materials on the occurrence of the Golden Ratio in the arts (e.g., the Parthenon, Egyptian temples, etc.). The books have been almost inexplicably influential.

Concerning "The Parade" specifically, while it is true that the horizontal is cut in proportions close to the Golden Ratio (in fact, the simple ratio eight-fifths), the vertical is not. An analysis of the entire composition of this and other paintings by Seurat, as well as paintings by the Symbolist painter Pierre Puvis de Chavannes (1824–1898), led even a Golden Ratio advocate like painter and author Charles Bouleau to conclude that "I do not think we can, without straining the evidence to regard his [Puvis de Chavannes's] compositions as based on the Golden Ratio. The same applies to Seurat." A detailed analysis in 1980 by Roger Herz-Fischler of all of Seurat's writings, sketches, and paintings reached the same conclusion. Furthermore, the mathematician, philosopher, and art critic Charles Henry (1859–1926) stated firmly in 1890 that the Golden Ratio was "perfectly ignored by contemporary artists."

Who, then, did use the Golden Ratio either in actual paintings or in the theory of painting? The first prominent artist and art theorist to employ the ratio was probably Paul Sérusier (1864–1927). Sérusier was born in Paris, and after studying philosophy he entered the famous art school Académie Julian. A meeting with the painters Paul Gaugin and Émile Bernard converted him to their expressive use of color and symbolist views. Together with the post-Impressionist painters Pierre Bonnard, Édouard Vuillard, Maurice Denis, and others he founded the group called the Nabis, from the Hebrew word meaning "prophets." The name was inspired by the group's half-serious, half-burlesque pose regarding their new style as a species of religious illumination. The composer Claude Debussy was also associated with the group. Sérusier probably heard about the Golden Ratio for the first time during one of his visits (between 1896 and 1903) to his friend the Dutch painter Jan Verkade (1868–1946). Verkade was a novice in the Benedictine monastery of Beuron, in South Germany. There groups of monk-painters were executing rather dull religious compositions based on "sacred measures," following a theory of Father Didier Lenz. According to Father Lenz's theory, the great art works of antiquity (e.g., Noah's Ark, Egyptian works, etc.) were all based on simple geometrical entities such as the circle, equilateral triangle, and hexagon. Sérusier found the charm of this theory captivating, and he wrote to Verkade: "as you can imagine, [I] have talked a great deal about your measures." The painter Maurice Denis (1870–1943) wrote biographical notes on Sérusier, from which we learn that those "measures" employed by Father Lenz included the Golden Ratio. Even though Sérusier admits that his initial studies of the mathematics of Beuron were "not all plain sailing," the Golden Ratio and the story of its potential association with the Great Pyramid and Greek artworks made it also into Sérusier's important art theory book *L'ABC de la Peinture* (The ABC of painting).

While Sérusier's interest in the Golden Ratio appears to have been more philosophical than practical, he did make use of this proportion in some of his works, mainly to "verify, and occasionally to check, his inventions of shapes and his composition."

Following Sérusier, the concept of the Golden Ratio propagated into other artistic circles, especially that of the Cubists. The name "Cubism" was coined by art critic Louis Vauxcelles (who, by the way, had also been responsible for "Expressionism" and "Fauvism") after viewing an exhibition of Georges Braque's work in 1908. The movement was inaugurated by Picasso's painting "Les Demoiselles d'Avignon" and Braque's "Nude." In revolt against the passionate use of color and form in Expressionism, Picasso and Braque developed an austere, almost monochrome style that deliberately rejected any subject matter that was likely to evoke emotional associations. Objects like musical instruments and even human figures were dissected into faceted geometrical planes, which were then combined in shifting perspectives. This analysis of solid forms for the purpose of revealing structure was quite amenable to the use of geometrical concepts like the Golden Ratio. In fact, some of the early Cubists, such as Jacques Villon and his brothers Marcel and Raymond Duchamp-Villon, together with Albert Gleizes and Francis Picabia, organized in Paris in October 1912 an entire exhibition entitled "Section d'Or" ("The Golden Section"). In spite of the highly suggestive name, none of the paintings that was exhibited actually included the Golden Section

Figure 77

as a basis for its composition. Rather, the organizers chose the name simply to project their general interest in questions that related art to science and philosophy. Nevertheless, some Cubists, like the Spanish-born painter Juan Gris (1887–1927) and the Lithuanian-born sculptor Jacques (Chaim Jacob) Lipchitz (1891–1973) did use the Golden Ratio in some of their later works. Lipchitz wrote: "At the time, I was very interested in theories of mathematical proportions, like the other cubists, and I tried to apply them to my sculptures. We all had a great curiosity for that idea of a golden rule or Golden Section, a system which was reputed to lay under the art and architecture of ancient Greece." Lipchitz helped Juan Gris in the construction of the sculpture "Arlequin" (currently in the Philadelphia Museum of Art; Figure 77), in which the two artists used Kepler's triangle (which is based on the Golden Ratio; see Figure 61) for the production of the desired proportions.

Another artist who used the Golden Ratio in the early 1920s was the Italian painter Gino Severini (1883–1966). Severini attempted in his work to reconcile the somewhat conflicting aims of Futurism and Cubism. Futurism represented an effort by a group of Italian intellectuals from literary arts, the visual arts, theater, music, and cinema to bring about a cultural rejuvenation in Italy. In Severini's words: "We choose to concentrate our attention on things in motion, because our modern sensibility is particularly qualified to grasp the idea of speed." The first painters' Futurist manifesto was signed in 1910, and it strongly urged the young Italian artists to "profoundly despise all forms of imitation." While still a Futurist himself, Severini found in Cubism a "notion of measure" that fit his ambition of "making, by means of painting, an object with the same perfection of craftsmanship as a cabinet maker making furniture." This striving for geometrical perfection led Severini to use the Golden Section in his preparatory drawings for several paintings (e.g., "Maternity," currently in a private collection in Rome; Figure 78).

Russian Cubist painter Maria Vorobëva, known as Marevna, provides an interesting instance of the role of the Golden Ratio in Cubist art. Marevna's 1974 book, *Life with the Painters of La Ruche,* is a fascinating account of the lives and works of her personal friends—a group that included the painters Picasso, Modigliani, Soutine, Rivera (with

Figure 78

whom she had a daughter), and others in Paris of the 1920s. Although Marevna does not give any specific examples and some of her historical comments are inaccurate, the text implies that Picasso, Rivera, and Gris had used the Golden Ratio as "another way of dividing planes, which is more complex and attracts experienced and inquisitive minds."

Another art theorist who had great interest in the Golden Ratio at the beginning of the twentieth century was the American Jay Hambidge (1867–1924). In a series of articles and books, Hambidge defined two types of symmetry in classical and modern art. One, which he called "static symmetry," was based on regular figures like the square and equilateral triangle, and was supposed to produce lifeless art. The other, which he dubbed "dynamic symmetry," had the Golden Ratio and the logarithmic spiral in leading roles. Hambidge's basic thesis was that the use of "dynamic symmetry" in design leads to vibrant and moving art. Few today take his ideas seriously.

One of the strongest advocates for the application of the Golden Ratio to art and architecture was the famous Swiss-French architect and painter Le Corbusier (Charles-Édouard Jeanneret, 1887–1965).

Jeanneret was born in La Chaux-de-Fonds, Switzerland, where he studied art and engraving. His father worked in the watch business as an enameler, while his mother was a pianist and music teacher who encouraged her son toward a musician's dexterity as well as more abstract pursuits. He began his studies of architecture in 1905 and eventually became one of the most influential figures in modern architecture. In the winter of 1916–1917, Jeanneret moved to Paris, where he met Amédée Ozenfant, who was well connected in the Parisian haut monde of artists and intellectuals. Through Ozenfant, Jeanneret met with the Cubists and was forced to grapple with their inheritance. In particular, he absorbed an interest in proportional systems and their role in aesthetics from Juan Gris. In the autumn of 1918, Jeanneret and Ozenfant exhibited together at the Galérie Thomas. More precisely, two canvases by Jeanneret were hung alongside many more paintings by Ozenfant. They called themselves "Purists," and entitled their catalog *Après le Cubisme* (After cubism). Purism invoked Piero della Francesca and the Platonic aesthetic theory in its assertion that "the work of art must not be accidental, exceptional, impressionistic, inorganic, protestatory, picturesque, but on the contrary, generalized, static, expressive of the invariant."

Jeanneret did not take the name "Le Corbusier" (co-opted from ancestors on his mother's side called Lecorbesier) until he was thirty-three, well installed in Paris, and confident of his future path. It was as if he wanted basically to repress his faltering first efforts and stimulate the myth that his architectural genius bloomed suddenly into full maturity.

Originally, Le Corbusier expressed rather skeptical, and even negative, views of the application of the Golden Ratio to art, warning against the "replacement of the mysticism of the sensibility by the Golden Section." In fact, a thorough analysis of Le Corbusier's architectural designs and "Purist" paintings by Roger Herz-Fischler shows that prior to 1927, Le Corbusier never used the Golden Ratio. This situation changed dramatically following the publication of Matila Ghyka's influential book *Aesthetics of Proportions in Nature and in the Arts,* and his *Golden Number, Pythagorean Rites and Rhythms* (1931) only enhanced the mystical aspects of φ even further. Le Corbusier's fascination with *Aesthetics* and with the Golden Ratio had two origins. On one hand, it was a consequence of his interest in basic forms and structures underlying natural phenomena. On the other, coming from a family that encouraged musical education, Le Corbusier could appreciate the Pythagorean craving for a harmony achieved by number ratios. He wrote: "More than these thirty years past, the sap of mathematics has flown through the veins of my work, both as an architect and painter; for music is always present within me." Le Corbusier's search for a standardized proportion culminated in the introduction of a new proportional system called the "Modulor."

The Modulor was supposed to provide "a harmonic measure to the human scale, universally applicable to architecture and mechanics." The latter quote is in fact no more than a rephrasing of Protagoras' famous saying from the fifth-century B.C. "Man is the measure of all things." Accordingly, in the spirit of the Vitruvian man (Figure 53) and the general philosophical commitment to discover a proportion system equivalent to that of natural creation, the Modulor was based on human proportions (Figure 79).

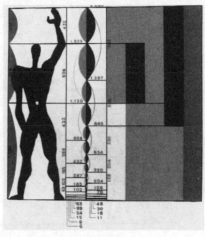

Figure 79

A six-foot (about 183-centimeter) man, somewhat resembling the familiar logo of the "Michelin man," with his arm upraised (to a height of 226 cm; 7'5"), was inserted into a square (Figure 80). The ratio of the height of the man (183 cm; 6') to the height of his navel (at the midpoint of 113 cm; 3' 8.5") was taken to be precisely in a Golden Ratio. The total height (from the feet to the raised arm) was also divided in a Golden Ratio (into 140 cm and 86 cm) at the level of the wrist of a downward-hanging arm. The two ratios (113/70) and (140/86) were further subdivided into smaller dimensions according to the Fibonacci series (each number being equal to the sum of the preceding two; Figure 81). In the final version of the Modulor (Figures 79 and 81), two scales of interspiraling Fibonacci dimensions were therefore introduced (the "red and the blue series").

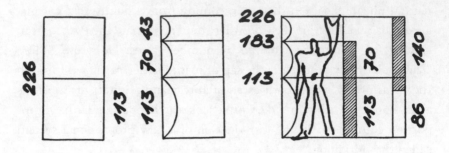

Figure 80

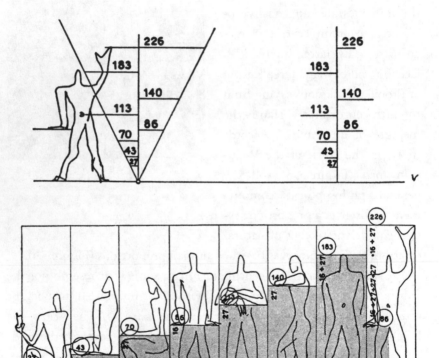

Figure 81

Le Corbusier suggested that the Modulor would give harmonious proportions to everything, from the sizes of cabinets and door handles, to buildings and urban spaces. In a world with an increasing need for mass production, the Modulor was supposed to provide the model for standardization. Le Corbusier's two books, *Le Modulor* (published in 1948) and *Modulor II* (1955), received very serious scholarly attention from architectural circles, and they continue to feature in any discussion of proportion. Le Corbusier was very proud of the fact that he had the opportunity to present the Modulor even to Albert Einstein, in a meeting at Princeton in 1946. In describing that event he says: "I expressed myself badly, I explained 'Modulor' badly, I got bogged down in the morass of 'cause and effect.' " Nevertheless, he received a letter from Einstein, in which the great man said this of the Modulor: "It is a scale of proportions which makes the bad difficult and the good easy."

Le Corbusier translated his theory of the Modulor into practice in many of his projects. For example, in his notes for the impressive urban layout of Chandigarh, India, which included four major government buildings—a Parliament, a High Court, and two museums—we find: "But, of course, the Modulor came in at the moment of partitioning the window area. . . . In the general section of the building which involves providing shelter from the sun for the offices and courts, the Modulor will bring textural unity in all places. In the design of the frontages, the Modulor (texturique) will apply its red and blue series within the spaces already furnished by the frames."

Figure 82

Le Corbusier was certainly not the last artist to be interested in the Golden Ratio, but most of those after him were fascinated more by the mathematical-philosophical-historical attributes of the ratio than by its presumed aesthetic properties. For example, the British abstract artist Anthony Hill used a Fibonacci series of dimensions in his 1960 "Constructional Relief" (Figure 82). Similarly, the contemporary Israeli painter and sculptor Igael Tumarkin has deliberately included the formula for the value of ϕ ($\phi = (1 + \sqrt{5})/2$) in one of his paintings.

An artist who transformed the Fibonacci sequence into an important ingredient of his art is the Italian Mario Merz. Merz was born in Milan in 1925, and in 1967 he joined the art movement labeled Arte Povera (Poor Art), which also included the artists Michelangelo Pistoletto, Luciano Fabro, and Jannis Kounellis. The name of the movement (coined by the critic Germano Celant) was derived from the desire of its members to use simple, everyday life materials, in a protest against what they regarded as a dehumanized, consumer-driven society. Merz started to use the Fibonacci sequence in 1970, in a series of "conceptual" works that include the numbers in the sequence or various spirals. Merz's desire to utilize Fibonacci numbers was based on the fact that the sequence underlies so many growth patterns of natural life. In a work from 1987 entitled "Onda d'urto" (Shock wave), he has a long row of stacks of newspapers, with the Fibonacci numbers glowing in blue neon lights above the stacks. The work "Fibonacci Naples" (from 1970) consists of ten photographs of factory workers, building in Fibonacci numbers from a solitary person to a group of fifty-five (the tenth Fibonacci number).

False claims about artists allegedly using the Golden Ratio continue to spring up almost like mushrooms after the rain. One of these claims deserves some special attention, since it is repeated endlessly.

The Dutch painter Piet Mondrian (1872–1944) is best known for his geometric, nonobjective style, which he called "neoplasticism." In particular, much of his art is characterized by compositions involving only vertical and horizontal lines, rectangles, and squares, and employing only primary colors (and sometimes black or grays) against a white background, as in "Broadway Boogie-Woogie" (Figure 83; in

The Museum of Modern Art, New York). Curved lines, three-dimensionality, and realistic representation were entirely eliminated from his work.

Not surprisingly, perhaps, Mondrian's geometrical compositions attracted quite a bit of Golden Numberist speculation. In *Mathematics,* David Bergamini admits that Mondrian himself "was vague about the design of his

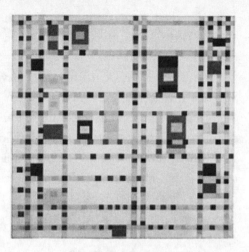

Figure 83

paintings," but nevertheless claims that the linear abstraction "Place de la Concorde" incorporates overlapping Golden Rectangles. Charles Bouleau was much bolder in *The Painter's Secret Geometry,* asserting that "the French painters never dared to go as far into pure geometry and the strict use of the golden section as did the cold and pitiless Dutchman Piet Mondrian." Bouleau further states that in "Broadway Boogie-Woogie," "the horizontals and verticals which make up this picture are nearly all in the golden ratio." With so many lines to choose from in this painting, it should come as no surprise that quite a few can be found at approximately the right separations. Having spent quite some time reading the more serious analyses of Mondrian's work and not having found any mention of the Golden Ratio there, I became quite intrigued by the question: Did Mondrian really use the Golden Ratio in his compositions or not? As a last resort I decided to turn to *the* real expert—Yves-Alain Bois of Harvard University, who coauthored the book *Mondrian* that accompanied the large retrospective exhibit of the artist's work in 1999. Bois's answer was quite categorical: "As far as I know, Mondrian never used a system of proportion (if one excepts the modular grids he painted in 1918–1919, but there the system is deduced from the format of the paintings themselves: they are divided in 8×8 units)." Bois added: "I also vaguely remember a remark by Mondrian mocking arithmetic computations with regard to his work." He concluded: "I think that the Golden Section is a complete red herring with regard to Mondrian."

All of this intricate history does leave us with a puzzling question. Short of intellectual curiosity, for what reason would so many artists even consider employing the Golden Ratio in their works? Does this ratio, as manifested for example in the Golden Rectangle, truly contain some intrinsic, aesthetically superior qualities? The attempts to answer this question alone resulted in a multitude of psychological experiments and a vast literature.

THE SENSES DELIGHT IN THINGS DULY PROPORTIONED

With the words in the title of this section, Italian scholastic philosopher St. Thomas Aquinas (ca. 1225–1274) attempted to capture a fundamental relationship between beauty and mathematics. Humans seem to react with a sense of pleasure to "forms" that possess certain symmetries or obey certain geometrical rules.

In our examination of the potential aesthetic value of the Golden Ratio, we will concentrate on the aesthetics of very simple, nonrepresentational forms and lines, not on complex visual materials and works of art. Furthermore, in most of the psychological experiments I shall describe, the term "beautiful" was actually shunned. Rather, words like "pleasing" or "attractive" have been used. This avoids the need for a definition of "beautiful" and builds on the fact that most people have a pretty good idea of what they like, even if they cannot quite explain why.

Numerous authors have claimed that the Golden Rectangle is the most aesthetically pleasing of all rectangles. The more modern interest in this question was largely initiated by a series of rather crankish publications by the German researcher Adolph Zeising, which started in 1854 with *Neue Lehre von den Proportionen des menschlichen Körpers* (The latest theory of proportions in the human body) and culminated in the publication (after Zeising's death) of a massive book, *Der Goldne Schnitt* (The golden section), in 1884. In these works, Zeising combined his own interpretation of Pythagorean and Vitruvian ideas to argue that "the partition of the human body, the structure of many animals which are characterized by well-developed building, the fundamental types of many forms of plants, . . . the harmonics of the most satisfying musical accords, and the proportionality of the most beautiful works in architecture and sculpture" are all based on the Golden Ratio. To him, therefore, the Golden Ratio offered the key to the understanding of all proportions in "the most refined forms of nature and art."

One of the founders of modern psychology, Gustav Theodor Fechner (1801–1887), took it upon himself to verify Zeising's pet theory. Fechner is considered a pioneer of experimental aesthetics. In one of his early experiments, he conducted a public opinion poll in which he asked all the visitors to the Dresden Gallery to compare the beauty of two nearly identical Madonna paintings (the "Darmstadt Madonna" and the "Dresden Madonna") that were exhibited together. Both paintings were attributed to the German painter Hans Holbein the Younger (1497–1543), but there was a suspicion that the "Dresden Madonna" was actually a later copy. That particular experiment resulted in a total failure—out of 11,842 visitors, only 113 answered the questionnaire, and even those were mostly art critics or people who had formed previous judgments.

Fechner's first experiments with rectangles were performed in the 1860s, and the results were published in the 1870s and eventually summarized in his 1876 book, *Vorschule der Aesthetik* (Introduction to aesthetics). He rebelled against a top-down approach to aesthetics, which starts with the formulation of abstract principles of beauty, and rather advocated the development of experimental aesthetics from the bottom up. The experiment was quite simple: Ten rectangles were placed in front of a subject who was asked to select the most pleasing one and the least pleasing one. The rectangles varied in their length-to-width ratios from a square (a ratio of 1.00) to an elongated rectangle (a ratio of 2.5). Three of the rectangles were more elongated than the Golden Rectangle, and six were closer to a square. According to Fechner's own description of the experimental setting, subjects often waited and wavered, rejecting one rectangle after another. Meanwhile the experimenter would explain that they should carefully select the most pleasing, harmonic, and elegant rectangle. In Fechner's experiment, 76 percent of all choices centered on the three rectangles having the ratios 1.75, 1.62, and 1.50, with the peak at the Golden Rectangle (1.62). All other rectangles received less than 10 percent of the choices each.

Fechner's motivation for studying the subject was not free of prejudice. He himself admitted that the inspiration for the research came to him when he "saw the vision of a unified world of thought, spirit and matter, linked together by the mystery of numbers." While nobody accuses Fechner of altering the results, some speculate that he may have subconsciously produced circumstances that would favor his desired outcome. In fact, Fechner's unpublished papers reveal that he conducted similar experiments with ellipses, and having failed to discover any preference for the Golden Ratio, he did not publish the results.

Fechner further measured the dimensions of thousands of printed books, picture frames in galleries, windows, and other rectangularly shaped objects. His results were quite interesting, and often amusing. For example, he found that German playing cards tended to be somewhat more elongated than the Golden Rectangle, while French playing cards were less so. On the other hand, he found the average height-to-width ratio of forty novels from the public library to be near ϕ. Paintings (the area inside the frame) were actually found to be "significantly shorter" than a Golden Rectangle. Fechner made the following (politically incorrect by today's standards) observation about window shapes: "Only the window shapes of the houses of peasants seem often to be square, which is consistent with the fact that people with a lower level of education prefer this form more than people with a higher education." Fechner further claimed that the point at which the transverse piece crosses the upright post in graveyard crosses divides the post, on the average, in a Golden Ratio.

Many researchers repeated similar experiments over the twentieth century, with varying results. Overly eager Golden Ratio enthusiasts usually report only those experiments that seem to support the idea of an aesthetic preference for the Golden Rectangle. However, more careful researchers point out the very crude nature and methodological defects of many of these experiments. Some found that the results depended, for example, on whether the rectangles were positioned with their long side horizontally or vertically, on the size and color of the rectangles, on the age of the subjects, on cultural differences, and especially on the experimental method used. In an article published in 1965, American psychologists L. A. Stone and L. G. Collins suggested that the preference for the Golden Rectangle indicated by some of the experiments was related to the area of the human visual field. These researchers found that an "average rectangle" of rectangles drawn within and around the binocular visual field of a variety of subjects has a length-to-width ratio of about 1.5, not too far from the Golden Ratio. Subsequent experiments, however, did not confirm Stone and Collins's speculation. In an experiment conducted in 1966 by H. R. Schiffman of Rutgers University, subjects were asked to "draw the most aesthetically pleasing rectangle" that they could on a sheet of paper. After completion, they were instructed to orient the figure either horizontally or vertically (with respect to the long side) in the most pleasing position. While Schiffman found an overwhelming preference for a horizontal orientation, consistent with the shape of the visual field, the average ratio of length to width was about 1.9—far from both the Golden Ratio and the visual field's "average rectangle."

The psychologist Michael Godkewitsch of the University of Toronto cast even greater doubts about the notion of the Golden Rectangle being the most pleasing rectangle. Godkewitsch first pointed out the important fact that average group preferences may not reflect at all the most preferred rectangle for each individual. Often something that is most preferred on the average is not chosen first by anyone. For example, the brand of chocolate that everybody rates second best may on the average be ranked as the best, but nobody will ever buy it! Consequently, first choices provide a more meaningful measure of preference than mean preference rankings. Godkewitsch further noted that if preference for the Golden Ratio is indeed universal and genuine, then it should receive the largest number of first choices, irrespective of which other rectangles the subjects are presented with.

Godkewitsch published in 1974 the results of a study that involved twenty-seven rectangles with length-to-width ratios in three ranges. In one range the Golden Rectangle was next to the most elongated rectangle, in one it was in the middle, and in the third it was next to the shortest rectangle. The results of the experiment showed, according to Godkewitsch, that the preference for the Golden Rectangle was an artifact of its position in the range of rectangles being presented and of the fact that mean preference rankings (rather than first choices) were used in the earlier experiments. Godkewitsch concluded that "the basic question whether there is or is not, in the Western world, a reliable verbally expressed *aesthetic* preference for a particular ratio between length and width of rectangular shapes can probably be answered negatively. Aesthetic theory has hardly any rationale left to regard the Golden Section as a decisive factor in formal visual beauty."

Not all agree with Godkewitsch's conclusions. British psychologist Chris McManus published in 1980 the results of a careful study that used the method of paired comparisons, whereby a judgment is made for each pair of rectangles. This method is considered to be superior to other experimental techniques, since there is good evidence that ranking tends to be a process of successive paired comparisons. McManus concluded that "there is moderately good evidence for the phenomenon which Fechner championed, even though Fechner's own method for its demonstration is, at best, highly suspect owing to methodological artifacts." McManus admitted, however, that "whether the Golden Section *per se* is important, as opposed to similar ratios (e.g. 1.5, 1.6 or even 1.75), is very unclear."

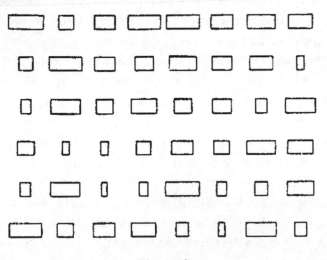

Figure 84

You can test yourself (or your friends) on the question of which rectangle you prefer best. Figure 84 shows a collection of forty-eight rectangles, all having the same height, but with their widths ranging from 0.4 to 2.5 times their height. University of Maine mathematician George Markowsky used this collection in his own informal experiments. Did you pick the Golden Rectangle as your first choice? (It is the fifth from the left in the fourth row.)

GOLDEN MUSIC

Every string quartet and symphony orchestra today still uses Pythagoras' discovery of whole-number relationships among the different musical tones. Furthermore, in the ancient Greek curriculum and up to medieval times, music was considered a part of mathematics, and musicians concentrated their efforts on the understanding of the mathematical basis of tones. The concept of the "music of the spheres" represented a glorious synthesis of music and mathematics, and in the imaginations of philosophers and musicians, it wove the entire cosmos into one grand design that could be perceived only by the gifted few. In the words of the great Roman orator and philosopher Cicero (ca. 106–43 B.C.): "The ears of mortals are filled with this sound, but they are unable to hear it. . . . You might as well try to stare directly at the Sun, whose rays are much too strong for your eyes." Only in the twelfth century did music break away from adherence to mathematical prescriptions and formulae. However, even as late as the eighteenth century, the German rationalist philosopher Gottfried Wilhelm Leibnitz (1646–1716) wrote: "Music is a secret arithmetical exercise and the person who indulges in it does not realize that he is manipulating numbers." Around the same time, the great German composer Johann Sebastian Bach (1685–1750) had a fascination for the kinds of games that can be played with musical notes and numbers. For example, he encrypted his signature in some of his compositions via musical codes. In the old German musical notation, B stood for B-flat and H stood for B-natural, so Bach could spell out his name in musical notes: B-flat, A, C, B-natural. Another encryption Bach used was based on Gematria.

Taking A = 1, B = 2, C = 3, and so on, B-A-C-H = 14 and
J-S-B-A-C-H = 41 (because I and J were the same letter in the German
alphabet of Bach's time). In his entertaining book *Bachanalia* (1994),
mathematician and Bach enthusiast Eric Altschuler gives numerous examples for the appearances of 14s (encoded BACH) and 41s (encoded
JSBACH) in Bach's music that he believes were put there deliberately
by Bach. For example, in the first fugue, the C Major Fugue, Book One
of Bach's *Well Tempered Clavier,* the subject has fourteen notes. Also, of
the twenty-four entries, twenty-two run all the way to completion and
a twenty-third runs almost all the way to completion. Only one entry—
the fourteenth—doesn't run anywhere near completion. Altschuler
speculates that Bach's obsession with encrypting his signature into his
compositions is similar to artists incorporating their own portraits into
their paintings or Alfred Hitchcock making a cameo appearance in each
of his movies.

Given this historical relationship between music and numbers, it is
only natural to wonder whether the Golden Ratio (and Fibonacci numbers) played any role either in the development of musical instruments
or in the composition of music.

The violin is an instrument in which the Golden Ratio does feature
frequently. Typically, the violin soundbox contains twelve or more arcs
of curvature (which make the violin's curves) on each side. The flat arc
at the base often is centered at the Golden Section point up the center
line.

Some of the best-known
violins were made by Antonio Stradivari (1644–1737)
of Cremona, Italy. Original
drawings (Figure 85) show
that Stradivari took special
care to place the "eyes" of the
f-holes geometrically, at positions determined by the
Golden Ratio. Few (if any)
believe that it is the application of the Golden Ratio that

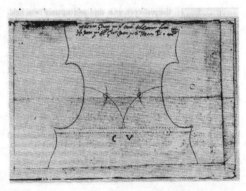

Figure 85

gives a Stradivarius violin its superior quality. More often such elements as varnish, sealer, wood, and general craftsmanship are cited as the potential "secret" ingredient. Many experts agree that the popularity of eighteenth-century violins in general stems from their adaptability for use in large concert halls. Most of these experts will also tell you that there is no "secret" in Stradivarius violins—these are simply inimitable works of art, the sum of all the parts that make up their superb craftsmanship.

Another musical instrument often mentioned in relation to Fibonacci numbers is the piano. The octave on a piano keyboard consists of thirteen keys, eight white keys and five black keys (Figure 86). The five black keys themselves form one group of two keys and another of three keys. The numbers 2, 3, 5, 8, and 13 happen all to be consecutive Fibonacci numbers. The primacy of the C major scale, for example, is partly due to the fact that it is being played on the piano's white keys.

However, the relationship between the piano keyboard and Fibonacci numbers is very probably a red herring. First, note that the chromatic scale (from C to B in the figure), which is fundamental to western music, is really composed of twelve, not thirteen, semitones. The same note, C, is played twice in the octave, to indicate the completion of the cycle. Second, and more important, the

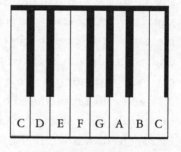

Figure 86

arrangement of the keys in two rows, with the sharp and flats being grouped in twos and threes in the upper row, dates back to the early fifteenth century, long before the publication of Pacioli's book and even longer before any serious understanding of Fibonacci numbers.

In the same way that Golden Numberists claim that the Golden Ratio has special aesthetic qualities in the visual arts, they also attribute to it particularly pleasing effects in music. For example, books on the Golden Ratio are quick to point out that many consider the major sixth and the minor sixth to be the most pleasing of musical intervals and that these intervals are related to the Golden Ratio. A pure musical tone is characterized by a fixed frequency (measured in the number of vibrations per second) and a fixed amplitude (which determines the instantaneous loudness). The standard tone used for tuning is A, which vibrates at 440 vibrations per second. A major sixth can be obtained from a combination of A with C, the latter note being produced by a frequency of about 264 vibrations per second. The ratio of the two frequencies 440/264 reduces to 5/3, the ratio of two Fibonacci numbers. A minor sixth can be obtained from a high C (528 vibrations per second) and an E (330 vibrations per second). The ratio in this case, 528/330, reduces to 8/5, which is also a ratio of two Fibonacci numbers and already very close to the Golden Ratio. (The ratios of successive Fibonacci numbers approach the Golden Ratio.) However, as in painting, note that in this case, too, the concept of a "most pleasing musical interval" is rather ambiguous.

Fixed-note instruments like the piano are tuned according to the "tempered scale" popularized by Bach, in which each semitone has an equal frequency ratio to the next semitone, making it easy to play in any key. The ratio of two adjacent frequencies in a well-tempered instrument is $2^{1/12}$ (the twelfth root of two). How was this number derived? Its origins actually can be traced to ancient Greece. Recall that an octave is obtained by dividing a string into two equal parts (a frequency ratio of 2:1), and a fifth is produced by a frequency ratio of 3:2 (basically using two-thirds of a string). One of the questions that intrigued the Pythagoreans was whether by repeating the procedure for creating the fifth (applying the 3/2 frequency ratio consecutively) one could generate an integer number of octaves. In mathematical terms, this means asking: Are there any two integers n and m such that $(3/2)^n$ is equal to 2^m? As it turns out, while no two integers satisfy this equality precisely, $n = 12$ and $m = 7$ come pretty close, because of the coincidence that $2^{1/12}$ is nearly equal to $3^{1/19}$ (the nineteenth root of 3). The twelve frequencies of the octave are therefore all approximate powers of the basic frequency ratio $2^{1/12}$. Incidentally, you may be amused to note that the ratio of 19/12 is equal to 1.58, not too far from ϕ.

Another way in which the Golden Ratio could, in principle, contribute to the satisfaction from a piece of music is through the concept of proportional balance. The situation here is somewhat trickier, however, than in the visual arts. A clumsily proportioned painting will instantly stick out in an exhibit like a sore thumb. In music, on the other hand, we have to hear the entire piece before making a judgment. Nevertheless, there is no question that experienced composers design the framework of their music in such a way that not only are the different parts in perfect balance with each other, but also each part in itself provides a fitting container for its musical argument.

We have seen many examples where Golden Ratio enthusiasts have scrutinized the proportions of numerous works in the visual arts to discover potential applications of φ. These aficionados have subjected many musical compositions to the same type of treatment. The results are very similar—alongside a few genuine utilizations of the Golden Ratio as a proportional system, there are many probable misconceptions.

Paul Larson of Temple University claimed in 1978 that he discovered the Golden Ratio in the earliest notated western music—the "Kyrie" chants from the collection of Gregorian chants known as Liber Usualis. The thirty Kyrie chants in the collection span a period of more than six hundred years, starting from the tenth century. Larson stated that he found a significant "event" (e.g., the beginning or ending of a musical phrase) at the Golden Ratio separation of 105 of the 146 sections of the Kyries he had analyzed. However, in the absence of any supporting historical justification or convincing rationale for the use of the Golden Ratio in these chants, I am afraid that this is no more than another exercise in number juggling.

In general, counting notes and pulses often reveals various numerical correlations between different sections of a musical work, and the analyst faces an understandable temptation to conclude that the composer introduced the numerical relationships. Yet, without a firmly documented basis (which is lacking in many cases), such assertions remain dubious.

In 1995, mathematician John F. Putz of Alma College in Michigan examined the question of whether Mozart (1756–1791) had used the Golden Ratio in the twenty-nine movements from his piano sonatas that consist of two distinct sections. Generally, these sonatas consist of two parts: the Exposition, in which the musical theme is first introduced, and the Development and Recapitulation, in which the main theme is further developed and revisited. Since musical pieces are divided into equal units of time called *measures* (or *bars*), Putz examined the ratios of the numbers of measures in the two sections of the sonatas. Mozart, who "talked of nothing, thought of nothing but figures" during his school days (according to his sister's testimony), is probably one of the better candidates for the use of mathematics in his compositions. In fact, several previous articles had claimed that Mozart's piano sonatas do reflect the Golden Ratio. Putz's first results appeared to be very promising. In the *Sonata No. 1 in C Major,* for example, the first movement consists of sixty-two measures in the Development and Recapitulation and thirty-eight in the Exposition. The ratio 62/38 = 1.63 is quite close to the Golden Ratio. However, a thorough examination of all the data basically convinced Putz that Mozart did *not* use the Golden Ratio in his sonatas, nor is it clear why the simple matter of measures would give a pleasing effect. It therefore appears that while many believe that Mozart's music is truly "divine," the "Divine Proportion" is not a part of it.

A famous composer who might have used the Golden Ratio quite extensively was the Hungarian Béla Bartók (1881–1945). A virtuoso pianist and folklorist, Bartók blended elements from other composers that he admired (including Strauss, Liszt, and Debussy) with folk music, to create his highly personal music. He once said that "the melodic world of my string quartets does not differ essentially from that of folk songs." The rhythmical vitality of his music, combined with a well-calculated formal symmetry, united to make him one of the most original twentieth-century composers.

The Hungarian musicologist Ernö Lendvai investigated Bartók's music painstakingly and published many books and articles on the subject. Lendvai testifies that "from stylistic analyses of Bartók's music I have been able to conclude that the chief feature of his chromatic technique is obedience to the laws of Golden Section in every movement."

According to Lendvai, Bartók's management of the rhythm of the composition provides an excellent example of his use of the Golden Ratio. By analyzing the fugue movement of Bartók's *Music for Strings, Percussion and Celesta,* for example, Lendvai shows that the eighty-nine measures of the movement are divided into two parts, one with fifty-five measures and the other with thirty-four measures, by the pyramid peak (in terms of loudness) of the movement. Further divisions are marked by the placement and removal of the sordini (the mutes for the instruments) and by other textural changes (Figure 87). All the numbers of measures are Fibonacci numbers, with the ratios between major parts (e.g., 55/34) being close to the Golden Ratio. Similarly, in *Sonata for Two Pianos and Percussion,* the various themes develop in Fibonacci/Golden Ratio order in terms of the numbers of semitones (Figure 88).

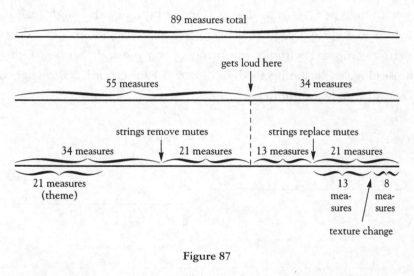

Figure 87

leitmotif	$3 + 5 = 8$
principal theme	$5 + 8 = 13$
secondary theme	13, 21

Figure 88

Some musicologists do not accept Lendvai's analyses. Lendvai himself admits that Bartók said nothing or very little about his own compositions, stating: "Let my music speak for itself; I lay no claim to any explanation of my works." The fact that Bartók did not leave any sketches to indicate that he derived rhythms or scales numerically makes any analysis suggestive at best. Also, Lendvai actually dodges the question of whether Bartók used the Golden Ratio consciously. Hungarian musicologist Laszlo Somfai totally discounts the notion that Bartók used the Golden Ratio, in his 1996 book *Béla Bartók: Composition, Concepts and Autograph Sources*. On the basis of a thorough analysis (which took three decades) of some 3,600 pages, Somfai concludes that Bartók composed without any preconceived musical theories. Other musicologists, including Ruth Tatlow and Paul Griffiths, also refer to Lendvai's study as "dubious."

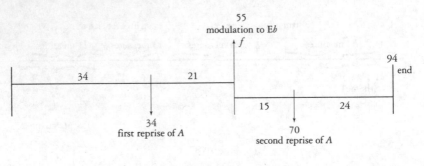

Figure 89

In the interesting book *Debussy in Proportion,* Roy Howat of Cambridge University argues that the French composer Claude Debussy (1862–1918), whose harmonic innovations had a profound influence on generations of composers, used the Golden Ratio in many of his compositions. For example, in the solo piano piece *Reflets dans l'eau* (Reflections in the water), a part of the series *Images,* the first rondo reprise occurs after bar 34, which is at the Golden Ratio point between the beginning of the piece and the onset of the climactic section after bar 55. Both 34 and 55 are, of course, Fibonacci numbers, and the ratio 34/21 is a good approximation for the Golden Ratio. The same structure is mirrored in the second part, which is divided in a 24/15 ratio (equal to the ratio of the two Fibonacci numbers 8/5, again close to the Golden Ratio; Figure 89). Howat finds similar divisions in the three symphonic sketches *La Mer* (The sea), in the piano piece *Jardins sous la Pluie* (Gardens under the rain), and other works.

I must admit that given the history of *La Mer,* I find it somewhat difficult to believe that Debussy used any mathematical design in the composition of this particular piece. He started *La Mer* in 1903, and in a letter he wrote to his friend André Messager he says: "You may not know that I was destined for a sailor's life and that it was only quite by chance that fate led me in another direction. But I have always retained a passionate love for her [the sea]." By the time the composition of *La Mer* was finished, in 1905, Debussy's whole life had been literally turned upside down. He had left his first wife, "Lily" (real name Rosalie Texier), for the alluring Emma Bardac; Lily attempted suicide; and both she and Bardac brought court actions against the composer. If you listen to *La Mer*—perhaps Debussy's most personal and passionate work—you can literally hear not only a musical portrait of the sea, probably inspired by the work of the English painter Joseph Mallord William Turner, but also an expression of the tumultuous period in the composer's life.

Since Debussy didn't say much about his compositional technique, we must maintain a clear distinction between what may be a forced interpretation imposed on the composition and the composer's real and conscious intention (which remains unknown). To support his analysis, Howat relies primarily on two pieces of circumstantial evidence: Debussy's close association with some of the symbolist painters who are known to have been interested in the Golden Ratio, and a letter Debussy wrote in August 1903 to his publisher, Jacque Durand. In that letter, which accompanied the corrected proofs of *Jardins sous la Pluie,* Debussy talks about a bar missing in the composition and explains: "However, it's necessary, as regards number; the divine number." The implication here is that not only was Debussy constructing his harmonic structure with numbers in general but that the "divine number" (assumed to refer to the Golden Ratio) played an important role.

Howat also suggests that Debussy was influenced by the writings of the mathematician and art critic Charles Henry, who had great interest in the numerical relationships inherent in melody, harmony, and rhythm. Henry's publications on aesthetics, such as the *Introduction à une esthétique scientifique* (Introduction to a scientific aesthetic; 1885), gave a prominent role to the Golden Ratio.

We shall probably never know with certainty whether this great pillar of French modernism truly intended to use the Golden Ratio to control formal proportions. One of his very few piano students, Mademoiselle Worms de Romilly, wrote once that he "always regretted not having worked at painting instead of music." Debussy's highly original musical aesthetic may have been aided, to a small degree, by the application of the Golden Ratio, but this was certainly not the main source of his creativity.

Just as a curiosity, the names of Debussy and Bartók are related through an amusing anecdote. During a visit of the young Hungarian composer to Paris, the great piano teacher Isidore Philipp offered to introduce Bartók to the composer Camille Saint-Saëns, at the time a great celebrity. Bartók declined. Philipp then offered him to meet with the great organist and composer Charles-Marie Widor. Again Bartók declined. "Well," said Philipp, "if you won't meet Saint-Saëns and Widor, who is there that you would like to know?" "Debussy," replied Bartók. "But he is a horrid man," said Philipp. "He hates everybody and will certainly be rude to you. Do you want to be insulted by Debussy?" "Yes," Bartók replied with no hesitation.

The introduction of recording technologies and computer music in the twentieth century accelerated precise numerical measurements and thereby encouraged number-based music. The Austrian composer Alban Berg (1885–1935), for example, constructed his Kammerkonzert entirely around the number 3: There are units of thirty bars, on three themes, with three basic "colors" (piano, violin, wind). The French composer Olivier Messiaen (1908–1992), who was largely driven by a deep Catholic faith and a love for nature, also used numbers consciously (e.g., to determine the number of movements) in rhythmic constructions. Nevertheless, when asked specifically in 1978 about the Golden Ratio, he disclaimed use of it.

The colorful composer, mathematician, and teacher Joseph Schillinger (1895–1943) exemplified by his own personality and teachings the Platonic view of the relationship between mathematics and music. After studying at the St. Petersburg Conservatory and teaching and composing at the Kharkov and Leningrad State academies, he settled in the United States in 1928, where he became a professor of both mathematics and music at various institutions, including Columbia University and New York University. The famous composer and pianist George Gershwin, the clarinetist and bandleader Benny Goodman, and the dance-band leader Glenn Miller were all among Schillinger's students. Schillinger was a great believer in the mathematical basis of music, and, in particular, he developed a System of Musical Composition in which successive notes in the melody followed Fibonacci intervals when counted in units of half-steps (Figure 90). To Schillinger, these Fibonacci leaps of the notes conveyed the same sense of harmony as the phyllotactic ratios of the leaves on a stem convey to the botanist. Schillinger found "music" in the most unusual places. In *Joseph Schillinger: A Memoir*, the biographical book written by his widow Frances, the author tells the story of a party riding in a car during a rain shower. Schillinger noted: "The splashing rain has its rhythm and the windshield wipers their rhythmic pattern. That's unconscious art." One of Schillinger's attempts to demonstrate that music can be based entirely on mathematical formulation was particularly amusing. He basically copied the fluctuations of a stock market curve as they appeared in the *New York Times* on graph paper and, by translating the ups and downs into proportional musical intervals, showed that he could obtain a composition somewhat similar to those of the great Johann Sebastian Bach.

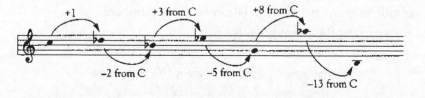

Figure 90

The conclusion from this brief tour of the world of music is that claims about certain composers having used the Golden Ratio in their music usually leap too swiftly from numbers generated by simple counting (of bars, notes, etc.) to interpretation. Nevertheless, there is no doubt that the twentieth century in particular produced a renewed interest in the use of numbers in music. As a part of this Pythagorean revival, the Golden Ratio also started to feature more prominently in the works of several composers.

The Viennese music critic Eduard Hanslick (1825–1904) expressed the relationship between music and mathematics magnificently in the book *The Beautiful in Music*:

The "music" of nature and the music of man belong to two distinct categories. The translation from the former to the latter passes through the science of mathematics. An important and pregnant proposition. Still, we should be wrong were we to construe it in the sense that man framed his musical system according to calculations purposely made, the system having arisen through the unconscious application of pre-existent conceptions of quantity and proportion, through subtle processes of measuring and counting; but the laws by which the latter are governed were demonstrated only subsequently by science.

PYTHAGORAS PLANNED IT

With the words in the heading, the famous Irish poet William Butler Yeats (1865–1939) starts his poem "The Statues." Yeats, who once stated that "the very essence of genius, of whatever kind, is precision," examines in the poem the relation between numbers and passion. The first stanza of the poem goes like this:

> *Pythagoras planned it. Why did the people stare?*
> *His numbers, though they moved or seemed to move*
> *In marble or in bronze, lacked character.*
> *But boys and girls, pale from the imagined love*

Of solitary beds, knew what they were,
That passion could bring character enough,
And pressed at midnight in some public place
Live lips upon a plummet-measured face.

Yeats emphasizes beautifully the fact that while the calculated proportions of Greek sculptures may seem cold to some, the young and passionate regarded these forms as the embodiment of the objects of their love.

At first glance, nothing seems more remote from mathematics than poetry. We think that the blossoming of a poem out of the poet's sheer imagination should be as boundless as the blossoming of a red rose. Yet recall that the growth of the rose's petals actually occurs in a well-orchestrated pattern based on the Golden Ratio. Could poetry be constructed on this basis, as well?

There are at least two ways, in principle, in which the Golden Ratio and Fibonacci numbers could be linked to poetry. First, there can be poems about the Golden Ratio or the Fibonacci numbers themselves (e.g., "Constantly Mean" by Paul Bruckman; presented in Chapter 4) or about geometrical shapes or phenomena that are closely related to the Golden Ratio. Second, there can be poems in which the Golden Ratio or Fibonacci numbers are somehow utilized in constructing the form, pattern, or rhythm.

Examples of the first type are provided by a humorous poem by J. A. Lindon, by Johann Wolfgang von Goethe's dramatic poem "Faust," and by Oliver Wendell Holmes's poem "The Chambered Nautilus."

Martin Gardner used Lindon's short poem to open the chapter on Fibonacci in his book *Mathematical Circus*. Referring to the recursive relation which defines the Fibonacci sequence, the poem reads:

Each wife of Fibonacci,
Eating nothing that wasn't starchy,
Weighed as much as the two before her,
His fifth was some signora!

Similarly, two lines from a poem by Katherine O'Brien read:

Fibonacci couldn't sleep—
Counted rabbits instead of sheep.

The German poet and dramatist Goethe (1743–1832) was certainly one of the greatest masters of world literature. His all-embracing genius is epitomized in *Faust*—a symbolic description of the human striving for knowledge and power. Faust, a learned German doctor, sells his soul to the devil (personified by Mephistopheles) in exchange for knowledge, youth, and magical power. When Mephistopheles finds that the pentagram's "Druidenfuss" ("Celtic wizard's foot") is drawn on Faust's threshold, he cannot get out. The magical powers attributed to the pentagram since the Pythagoreans (and which led to the definition of the Golden Ratio) gained additional symbolic meaning in Christianity, since the five vertices were assumed to stand for the letters in the name Jesus. As such, the pentagram was taken to be a source of fear for the devil. The text reads:

> *Mephistopheles:* Let me admit; a tiny obstacle
> Forbids my walking out of here:
> It is the druid's foot upon your threshold.
> *Faust:* The pentagram distresses you?
> But tell me, then, you son of hell.
> If this impedes you, how did you come in?
> *Mephistopheles:* Observe! The lines are poorly drawn;
> That one, the angle pointing outward,
> Is, you see, a little open.

Mephistopheles therefore uses trickery—the fact that the pentagram had a small opening in it—to get by. Clearly, Goethe had no intention of referring to the mathematical concept of the Golden Ratio in *Faust,* and he included the pentagram only for its symbolic qualities. Goethe expressed elsewhere his opinion on mathematics thus: "The mathematicians are a sort of Frenchmen: when you talk to them, they immediately translate it into their own language, and right away it is something entirely different."

The American physician and author Oliver Wendell Holmes

(1809–1894) published a few collections of witty and charming poems. In "The Chambered Nautilus" he finds a moral in the self-similar growth of the logarithmic spiral that characterizes the mollusk's shell:

> Build thee more stately mansions, O my soul,
>   As the swift seasons roll!
> Leave thy low-vaulted past!
>   Let each new temple, nobler than the last,
> Shut thee from heaven with a dome more vast,
>   Till thou at length art free,
> Leaving thine outgrown shell by life's unresting sea.

There are many examples of numerically based poetic structures. For example, the *Divine Comedy,* the colossal literary classic by the Italian poet Dante Alighieri (1265–1321), is divided into three parts, written in units of three lines, and each of the parts has thirty-three cantos (except for the first, which has thirty-four cantos, to bring the total to an even one hundred).

Poetry is probably the place in which Fibonacci numbers made their first appearance, even before Fibonacci's rabbits. One of the categories of meters in Sanskrit and Prakit poetry is known as mātrā-vittas. These are meters in which the number of morae (ordinary short syllables) remains constant and the number of letters is arbitrary. In 1985, mathematician Parmanand Singh of Raj Norain College, India, pointed out that Fibonacci numbers and the relation that defines them appeared in the writings of three Indian authorities on mātrā-vittas before A.D. 1202, the year in which Fibonacci's book was published. The first of these authors on metric was Ācārya Virahānka, who lived sometime between the sixth and eighth centuries. Although the rule he gives is somewhat vague, he does mention mixing the variations of two earlier meters to obtain the next one, just as each Fibonacci number is the sum of the two preceding ones. The second author, Gopāla, gives the rule specifically in a manuscript written between 1133 and 1135. He explains that each meter is the sum of the two earlier meters and calculates the series of meters 1, 2, 3, 5, 8, 13, 21 . . . , which is precisely the Fibonacci sequence. Finally, the great Jain writer Ācārya Hemacandra, who lived in the twelfth century and enjoyed the patronage of two kings, also stated clearly in a manuscript written around 1150 that "sum of the last and the last but one numbers [of variations] is [that] of the mātrā-vitta coming next." However, these early poetic appearances of Fibonacci numbers went apparently unnoticed by mathematicians.

In her educational book *Fascinating Fibonaccis,* author Trudi Hammel Garland gives an example of a limerick in which the number of lines (5), the number of beats in each line (2 or 3), and the total number of beats (13) are all Fibonacci numbers.

> *A fly and a flea in a flue (3 beats)*
> *Were imprisoned, so what could they do? (3 beats)*
> *Said the fly, "Let us flee!" (2 beats)*
> *"Let us fly!" said the flea, (2 beats)*
> *So they fled through a flaw in the flue. (3 beats)*

We should not take the appearance of very few Fibonacci numbers as evidence that the poet necessarily had these numbers or the Golden Ratio in mind when constructing the structural pattern of the poem. Like music, poetry is, and especially was, often intended to be heard, not just read. Consequently, proportion and harmony that appeal to the ear are an important structural element. This does not mean, however, that the Golden Ratio or Fibonacci numbers are the only options in the poet's arsenal.

George Eckel Duckworth, a professor of classics at Princeton University, made the most dramatic claim about the appearance of the Golden Ratio in poetry. In his 1962 book *Structural Patterns and Proportions in Vergil's Aeneid,* Duckworth states that "Vergil composed the *Aeneid* on the basis of mathematical proportion; each book reveals, in small units as well as in the main divisions, the famous numerical ratio known variously as the Golden Section, the Divine Proportion, or the Golden Mean ratio."

The Roman poet Vergil (70 B.C.–19 B.C.) grew up on a farm, and many of his early pastoral poems deal with the charm of rural life. His national epic the *Aeneid,* which details the adventures of the Trojan hero Aeneas, is considered one of the greatest poetic works in history. In twelve books, Vergil follows Aeneas from his escape from Troy to Carthage, through his love affair with Dido, to the establishment of the Roman state. Vergil makes Aeneas the paragon of piety, devotion to family, and loyalty to state.

Duckworth made detailed measurements of the lengths of passages in the *Aeneid* and computed the ratios of these lengths. Specifically, he measured the number of lines in passages characterized as major (and denoted that number by M) and minor (and denoted the number by m), and calculated the ratios of these numbers. The identification of major and minor parts was based on content. For example, in many passages the major or minor part is a speech and the other part (minor or major respectively) is a narrative or a description. From this analysis Duckworth concluded that the *Aeneid* contains "hundreds of Golden Mean ratios." He also noted that an earlier analysis (from 1949) of another Vergil work *(Georgius I)* gave for the ratio of the two parts (in terms of numbers of lines), known as "Works" and "Days," a value very close to ϕ.

Unfortunately, Roger Herz-Fischler has shown that Duckworth's analysis probably is based on a mathematical misunderstanding. Since this oversight is endemic to many of the "discoveries" of the Golden Ratio, I will explain it here briefly.

Suppose you have any pair of positive values m and M, such that M is larger than m. For example, $M = 317$ could be the number of pages in the last book you read and $m = 160$ could be your weight in pounds. We could represent these two numbers on a line (with proportional lengths), as in Figure 91. The ratio of the shorter to the longer part is equal to $m/M = 160/317 = 0.504$, while the ratio of the longer part to the whole is $M/(M + m) = 317/477 = 0.665$. You will notice that the value of $M/(M + m)$ is closer to $1/\phi = 0.618$ than m/M. We can prove mathematically that this is always the case. (Try it with the actual number of pages in your last book and your real weight.) From the definition of the Golden Ratio, we know that when a line is divided in a Golden Ratio, $m/M = M/(M + m)$ precisely. Consequently, we may be tempted to think that if we examine a series of ratios of numbers, such as the lengths of passages, for the potential presence of the Golden Ratio, it does not matter if we look at the ratio of the shorter to the longer or the longer to the whole. What I have just shown is that it definitely does matter. A too-eager Golden Ratio enthusiast wishing to demonstrate a Golden Ratio relationship between the weights of readers and the numbers of pages in the books they read may be able to do so by presenting data in the form $M/(M + m)$, which is biased toward $1/\phi$. This is precisely what happened to Duckworth. By making the unfortunate decision to use only the ratio $M/(M + m)$ in his analysis, because he thought that this was "slightly more accurate," he compressed and distorted the data and made the analysis statistically invalid. In fact, Leonard A. Curchin of the University of Ottawa and Roger Herz-Fischler repeated in 1981 the analysis with Duckworth's data (but using the ratio m/M) and showed that there is no evidence for the Golden Ratio in the *Aeneid.* Rather, they concluded that "random scattering is indeed the case with Vergil." Furthermore, Duckworth "endowed" Vergil with the knowledge that the ratio of two consecutive Fibonacci numbers is a good approximation of the Golden Ratio. Curchin and Herz-Fischler, on the other hand, demonstrated convincingly that even Hero of Alexandria, who lived later than Vergil and was one of the distinguished mathematicians of his time, did not know about this relation between the Golden Ratio and Fibonacci numbers.

$M = 317$ $m = 160$

Figure 91

Sadly, the statement about Vergil and ϕ continues to feature in most of the Golden Ratio literature, again demonstrating the power of Golden Numberism.

All the attempts to disclose the (real or false) Golden Ratio in various works of art, pieces of music, or poetry rely on the assumption that a canon for ideal beauty exists and can be turned to practical account. History has shown, however, that the artists who have produced works of lasting value are precisely those who have broken away from such academic precepts. In spite of the Golden Ratio's importance for many areas of mathematics, the sciences, and natural phenomena, we should, in my humble opinion, give up its application as a fixed standard for aesthetics, either in the human form or as a touchstone for the fine arts.

8

FROM THE TILES TO
THE HEAVENS

*Understanding is, after all, what science is all about—and science
is a great deal more than mere mindless computation.*
—ROGER PENROSE (1931–)

The tangled tale of the Golden Ratio has taken us from the sixth century B.C. to contemporary times. Two intertwined trends thread these twenty-six centuries of history. On one hand, the Pythagorean motto "all is number" has materialized spectacularly, in the role that the Golden Ratio plays in natural phenomena ranging from phyllotaxis to the shape of galaxies. On the other, the Pythagorean obsession with the symbolic meaning of the pentagon has metamorphosed into what I believe is the false notion that the Golden Ratio provides a universal canon of ideal beauty. After all of this, you may wonder whether there still is room left for any further exploration of this seemingly simple division of a line.

THE TILED ROAD TO QUASI-CRYSTALS

The Dutch painter Johannes Vermeer (1632–1675) is best known for his fantastically alluring genre paintings, which typically show one or

Figure 92   Figure 93

two figures engaged in some domestic task. In many of these paintings, a window on the viewer's left softly lights the room, and the way the light reflects off the tiled floor is purely magical. If you examine some of these paintings closely, you will find that quite a few, such as "The Concert," "A Lady Writing a Letter with Her Maid," "Love Letter" (Figure 92; located in the Rijksmuseum, Amsterdam), and "The Art of Painting" (Figure 93; located in the Kunsthistorisches Museum, Vienna), have identical floor tiling patterns, composed of black and white squares.

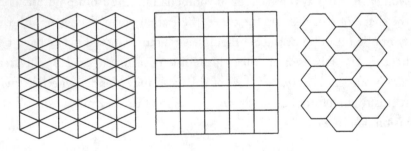

Figure 94

Squares, equilateral triangles, and hexagons are particularly easy to tile with, if one wants to cover the entire plane and achieve a pattern that repeats itself at regular intervals—known as periodic tiling (Figure

94). Simple, undecorated square tiles and the patterns they form have a fourfold symmetry—when rotated through a quarter of a circle (90 degrees), they remain the same. Similarly, equilateral, triangular tiles have a threefold symmetry (they remain the same when rotated by a third of a circle or 120 degrees), and hexagonal tiles have a sixfold symmetry (they remain the same when rotated by 60 degrees).

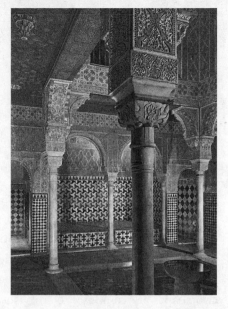

Figure 95

Periodic tilings also can be generated with more complex shapes. One of the most astounding monuments of Islamic architecture, the citadel-palace Alhambra in Granada, Spain, contains numerous examples of intricate tilings (Figure 95). Some of those patterns inspired the famous Dutch graphic artist M. C. Escher (1898–1972), who produced many imaginative examples of tilings (e.g., Figure 96), to which he referred as "divisions of the plane."

The geometrical plane figure most directly related to the Golden Ratio is, of course, the regular pentagon, which has a fivefold symmetry. Pentagons, however, cannot be used to fill the plane entirely and form a periodic tiling pattern. No matter how hard you try, unfilled gaps will remain. Consequently, it has long been thought that no tiling pattern

Figure 96

with long-range order can also exhibit a fivefold symmetry. However, in 1974, Roger Penrose discovered two basic sets of tiles that can fit together to fill the entire plane and exhibit the "forbidden" five-fold rotational symmetry. The resulting patterns are not strictly periodic, even though they display a long-range order.

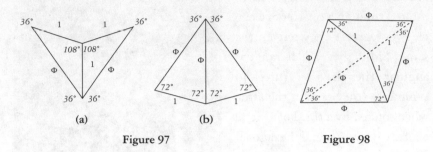

(a)   (b)

Figure 97   **Figure 98**

The Penrose tilings have the Golden Ratio written all over them. One pair of tiles that Penrose considered consists of two shapes known as a "dart" and a "kite" (Figure 97; a and b, respectively). Note that the two shapes are composed of the isosceles triangles that appear in the pentagon (Figure 25). The triangle in which the ratio of side to base is ϕ (Figure 97b) is the one known as a Golden Triangle, and the one in which the ratio of side to base is $1/\phi$ (Figure 97a) is the one known as a Golden Gnomon. The two shapes can be obtained by cutting a diamond shape or rhombus with angles of 72 degrees and 108 degrees in a way that divides the long diagonal in a Golden Ratio (Figure 98).

Penrose and Princeton mathematician John Horton Conway showed that in order to cover the whole plane with darts and kites in a nonperiodic way (as in Figure 99), certain matching rules must be obeyed. The latter can be ensured by adding "keys" in the form of notches and protrusions on the sides of the figures, like in the pieces of a jigsaw puzzle (Figure 100). Penrose and Conway further

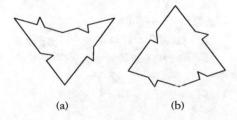

(a)   (b)

Figure 99   **Figure 100**

proved that darts and kites can fill the plane in infinitely many nonperiodic ways, with every pattern that can be discerned being surrounded by every other pattern. One of the most startling properties of any Penrose kite-dart tiling design is that the number of kites is about 1.618 times the number of darts. That is, if we denote by N_{kites} the number of kites and N_{darts} the number of darts, then N_{kites}/N_{darts} approaches ϕ the larger the area we take in.

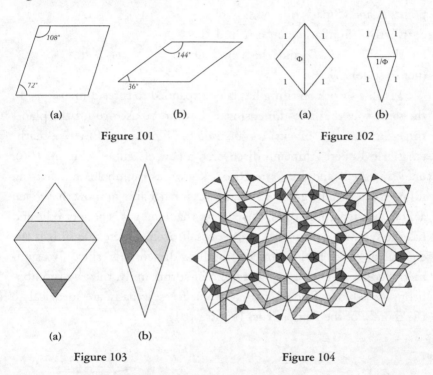

Figure 101   Figure 102

Figure 103   Figure 104

Another pair of Penrose tiles that can fill the entire plane (nonperiodically) is composed of two diamonds (rhombi), one fat (obtuse) and one thin (acute; Figure 101). As in the kite-dart pair, each of the rhombi is composed to two Golden Triangles or Golden Gnomons (Figure 102), and special matching rules have to be obeyed (in this case described by decorating the appropriate sides or angles of the rhombi; Figure 103) to obtain a plane-filling pattern (as in Figure 104). Again, in large areas there are 1.618 times more fat rhombi than thin ones, $N_{fat}/N_{thin} = \phi$.

The fat and thin rhombi are intimately related to the darts and kites and both, through the Golden Ratio, to the pentagon-pentagram sys-

tem. Recall that the Pythagorean interest in the Golden Ratio started with the
infinite series of nested pentagons and
pentagrams in Figure 105. All four of
the Penrose tiles are hidden in this figure. Points *B* and *D* mark the opposite
far corners of the kite *DCBA,* while
points *A* and *C* mark the "wings" of the
dart *EABC.* Similarly, you can find the
fat rhombus *AECD* and the thin one
(not to scale) *ABCF.*

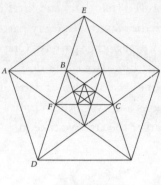

Figure 105

Penrose's work on tiling has been expanded to three dimensions. In
the same way that two-dimensional tiles can be used to fill the plane,
three-dimensional "blocks" can be used to fill up space. In 1976, mathematician Robert Ammann discovered a pair of "cubes" (Figure 106),
one "squashed" and one "stretched," known as rhombohedra, that can
fill up space with no gaps. Ammann was further able to show that given
a set of face-matching rules, the pattern that emerges is nonperiodic and
has the symmetry properties of the icosahedron (Figure 20e; this is the
equivalent of fivefold symmetry in three dimensions, since five symmetric edges meet at every vertex). Not surprisingly, the two rhombohedra are Golden Rhombohedra—their faces actually are identical to
the rhombi of the Penrose tiles (Figure 101).

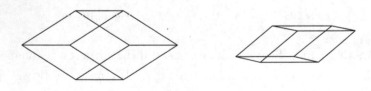

Figure 106

Penrose's tilings might have remained in the relative obscurity of
recreational mathematics were it not for a dramatic discovery in 1984.
Israeli materials engineer Dany Schectman and his collaborators found
that the crystals of an aluminum manganese alloy exhibit both long-
range order and fivefold symmetry. This was just about as shocking to

crystallographers as the discovery of a herd of five-legged cows would be to zoologists. For decades, solid-state physicists and crystallographers were convinced that solids can come in only two basic forms: Either they are highly ordered and fully periodic crystals, or they are totally amorphous. In ordered crystals, like those of ordinary table salt, atoms or groups of atoms appear in precisely recurring motifs, called *unit cells,* which form periodic structures. For example, in salt, the unit cell is a cube, and each chlorine atom is surrounded by sodium neighbors and vice versa (Figure 107). Just as in a perfectly tiled floor, the position and orientation of each unit cell determines uniquely the entire pattern. In amorphous materials, such as glasses, on the other hand, the atoms are totally disordered. In the same way that only shapes like squares (with a fourfold symmetry), triangles (threefold symmetry), and hexagons (sixfold symmetry) can fill the entire plane with a periodic tiling, only crystals with two-, three-, four-, and six-fold symmetry were thought to exist. Schect-man's crystals caused complete bewilderment because they appeared both to be highly or-

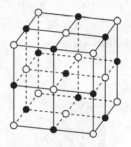

Figure 107

dered (like periodic crystals) and to exhibit fivefold (or icosahedral) symmetry. Before this discovery, few people suspected that another state of matter could exist, sharing important aspects with both crystalline and amorphous substances.

These new kinds of crystals (since the original discovery, other alloys of aluminum have been found) are now known as *quasi-crystals*—they are neither amorphous like glass nor precisely periodic like salt. In other words, these unusual materials appear to have precisely the properties of Penrose tilings! But this realization by itself is of little use to physicists, who want to understand why and how the quasi-crystals form. Penrose's and Ammann's matching rules are in this case little more than a clever mathematical exercise that does not explain the behavior of real atoms or atom clusters. In particular, it is difficult to imagine energetics that permit precisely the existence of two types of clusters (like the two Ammann rhombohedra) in just the required proportion in terms of density.

A clue toward a possible explanation came in 1991, when mathematician Sergei E. Burkov of the Landau Institute of Theoretical
Physics in Moscow realized that two shapes of tiles are not needed to
achieve quasi-periodic tiling in the plane. Burkov showed that quasi-
periodicity could be generated even using a single, decagonal (ten-
sided) unit, provided that the tiles are allowed to overlap—a property
forbidden in previous tiling attempts. Five years later, German mathematician Petra Gummelt of the Ernst Moritz Arndt University in
Greifswald proved rigorously that Penrose tiling can be obtained by using a single "decorated" decagon combined with a specific overlapping
rule. Two decagons may overlap only if shaded areas in the decoration
overlap (Figure 108). The decagon is also closely related to the Golden
Ratio—the radius of the circle that circumscribes a decagon with a side
length of 1 unit is equal to ϕ.

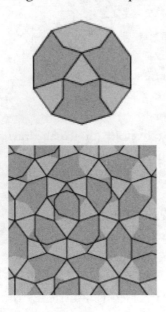

Based on Gummelt's work, mathematics finally could be turned into physics.
Physicists Paul Steinhardt of Princeton
University and Hyeong-Chai Jeong of Se-
jong University in Seoul showed that the
purely mathematical rules of overlapping
units could be transformed into a physical
picture in which "quasi-unit cells," which
are really clusters of atoms, simply share
atoms. Steinhardt and Jeong suggested that
quasi-crystals are structures in which identical clusters of atoms (quasi-unit cells)
share atoms with their neighbors, in a pattern that is designed to maximize the clus-
ter density. In other words, quasi-periodic
packing produces a system that is more stable (higher density and lower energy) than

Figure 108

otherwise. Steinhardt, Jeong, and collaborators also attempted to verify
this model experimentally in 1998. They bombarded a quasi-crystal alloy of aluminum, nickel, and cobalt with X-ray and electron beams.
The images of the structure obtained from the scattered beams were in

remarkable agreement with the picture of overlapping decagons. This is shown in Figure 109, where a decagon tiling pattern is superimposed on the experimental result. More recent experiments gave results that were somewhat more ambiguous. Nevertheless, the general impression remains that quasi-crystals can be explained by the Steinhardt-Jeong model.

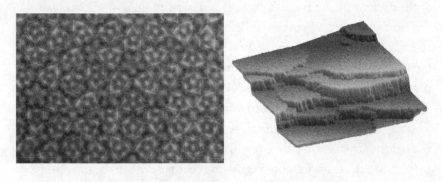

Figure 109   Figure 110

Images of the surfaces of quasi-crystals (taken in 1994 and 2001) reveal another fascinating relation to the Golden Ratio. Using a technique known as scanning tunneling microscopy (STM), scientists from the University of Basel, Switzerland, and from the Ames Laboratory at Iowa State University were able to obtain high-resolution images of the surfaces of an aluminum-copper-iron alloy and an aluminum-palladium-manganese alloy, both of which are quasi-crystals. The images show flat "terraces" (Figure 110) terminating in steps that come primarily in two heights, "high" and "low" (both measuring only a few hundred-millionths of an inch). The ratio of the two heights was found to be equal to the Golden Ratio!

Quasi-crystals are a magnificent example of a concept that started out as a purely mathematical entity (based on the Golden Ratio) but that eventually provided an explanation of a real, natural phenomenon. What is even more amazing about this particular development is that the concept emerged out of *recreational* mathematics. How could mathematicians have "anticipated" later discoveries by physicists? The question becomes more intriguing yet when we recall that Dürer and Kepler showed interest in tilings with fivefold symmetric shapes already in the sixteenth and seventeenth centuries. Can even the most esoteric topics in mathematics eventually find applications in either natural or human-inspired phenomena? We shall return to this question in Chapter 9.

Another fascinating aspect of the quasi-crystals story is related to two of the main theorists involved. Both Penrose and Steinhardt spent much of their scientific careers on topics related to cosmology—the study of the universe as a whole. Penrose is the person who discovered that Einstein's theory of general relativity predicts its own defects, points in which the strength of gravity becomes infinite. These mathematical singularities correspond to the objects we call *black holes,* which are masses that have collapsed to such densities that their gravity is sufficiently strong to prevent any light, mass, or energy to escape from them. Observations during the past quarter century have revealed that black holes are not just imaginary theoretical concepts but actual objects that exist in the universe. Recent observations with the two large space observatories, the Hubble Space Telescope and the Chandra X-ray Observatory, have shown that black holes are not even very rare. Rather, the centers of most galaxies harbor monstrous black holes with masses between a few million and a few billion times the mass of our Sun. The presence of the black holes is revealed by the gravitational pull they exert on stars and gas in their neighborhood. According to the standard "big bang" model that describes the origin of our entire universe, the cosmos as a whole started its expansion from such a singularity—an extremely hot and dense state.

Paul Steinhardt was one of the key figures in the development of what is known as the inflationary model of the universe. According to this model, originally proposed by physicist Alan Guth of MIT, when the universe was only a tiny fraction of a second old (0.000 . . . 1; with the "1" at the 35th decimal place), it underwent a fantastically rapid expansion, increasing in size by a factor of more than 10^{30} (1 followed by 30 zeros) within a fraction of a second. This model explains a few otherwise puzzling properties of our universe, such as the fact that it looks almost precisely the same in every direction—it is exquisitely isotropic. In 2001, Steinhardt and collaborators proposed a new version of the universe's very beginnings, known as the Ekpyrotic Universe (from the

Greek word for "conflagration," or a sudden burst of fire). In this still very speculative model, the big bang occurred when two three-dimensional universes moving along a hidden extra dimension collided.

The intriguing question is: Why did these two outstanding cosmologists decide to get involved in recreational mathematics and quasicrystals?

I have known Penrose and Steinhardt for many years, being in the same business of theoretical astrophysics and cosmology. In fact, Penrose was an invited speaker in the first large conference that I organized on relativistic astrophysics in 1984, and Steinhardt was an invited speaker in the latest one in 2001. Still, I did not know what motivated them to delve into recreational mathematics, which appears to be rather remote from their professional interests in astrophysics, so I asked them.

Roger Penrose replied: "I am not sure I have a deep answer for that. As you know, mathematics is something most mathematicians do for enjoyment." After some reflection he added: "I used to play with shapes fitting together since I was a child; some of my work on tiles therefore predated my work in cosmology. At the particular time, however, my recreational mathematics work was at least partially motivated by my cosmological research. I was thinking about the large-scale structure of the universe and was looking for toy models with simple basic rules, which could nevertheless generate complicated structures on large scales."

"But," I asked, "what was it that induced you to continue to work on that problem for quite a while?"

Penrose laughed and said, "As you know, I have always been interested in geometry; that problem simply intrigued me. Furthermore, while I had a hunch that such structures could occur in nature, I just couldn't see how nature could assemble them through the normal process of crystal growth, which is local. To some extent I am still puzzled by that."

Paul Steinhardt's immediate reaction on the phone was: "Good question!" After thinking about it for a few minutes he reminisced: "As an undergraduate student I really wasn't sure what I wanted to do. Then, in graduate school, I looked for some mental relief from my

strenuous efforts in particle physics, and I found that in the topic of order and symmetry in solids. Once I stumbled on the problem of quasi-
periodic crystals, I found it *irresistible* and I kept coming back to it."

FRACTALS

The Steinhardt-Jeong model for quasi-crystals has the interesting property that it produces long-range order from neighborly interactions,
without resulting in a fully periodic crystal. Amazingly enough, we can
also find this general property in the Fibonacci sequence. Consider the
following simple algorithm for the creation of a sequence known as the
Golden Sequence. Start with the number 1, and then replace 1 by 10.
From then on, replace each 1 by 10 and each 0 by 1. You will obtain the
following steps:

1

10

101

10110

10110101

1011010110110

101101011011010110101

and so on. Clearly, we started here with a "short-range" law (the simple
transformation of $0 \rightarrow 1$ and $1 \rightarrow 10$) and obtained a nonperiodic long-
range order. Note that the numbers of 1s in the sequence of lines 1, 1,
2, 3, 5, 8 . . . form a Fibonacci sequence, and so do the numbers of 0s
(starting from the second line). Furthermore, the ratio of the number of
1s to the number of 0s approaches the Golden Ratio as the sequence
lengthens. In fact, an examination of Figure 27 reveals that if we take 0
to stand for a baby pair of rabbits and 1 to stand for a mature pair, then
the sequence just given mirrors precisely the numbers of rabbit pairs.
But there is even more to the Golden Sequence than these surprising
properties. By starting with 1 (on the first line), followed by 10 (on the
second line), and simply appending to each line the line just preceding

it, we can also generate the entire sequence. For example, the fourth line, 10110, is obtained by appending the second line, 10, to the third, 101, and so on.

Recall that "self-similarity" means symmetry across size scale. The logarithmic spiral displays self-similarity because it looks precisely the same under any magnification, and so does the series of nested pentagons and pentagrams in Figure 10. Every time you walk into a hair stylist shop, you see an infinite series of self-similar reflections of yourself between two parallel mirrors.

The Golden Sequence is also self-similar on different scales. Take the sequence

$$1\,0\,1\,1\,0\,1\,0\,1\,1\,0\,1\,1\,0\,1\,0\,1\,1\ldots$$

and probe it with a magnifying glass in the following sense. Starting from the left, whenever you encounter a 1, mark a group of three symbols, and when you encounter a 0, mark a group of two symbols (with no overlap among the different groups). For example, the first digit is a 1, we therefore mark the group of the first three digits 101 (see below). The second digit from the left is a zero, therefore we mark the group of two digits 10 that *follow* the first 101. The third digit is 1; therefore we mark the three digits 101 that follow the 10; and so on. The marked sequence now looks like this

$$\overgroup{101}\ \overgroup{10}\ \overgroup{101}\ \overgroup{101}\ \overgroup{10}\ \overgroup{101}\ldots$$

Now from every group of three symbols retain the first two, and from every group of two retain the first one (the retained symbols are underlined):

$$\underline{10}\,1\,\underline{10}\,1\,0\,\underline{10}\,1\,\underline{10}\,1\,1\,0\,1\,0\,\underline{10}\,1\,1\ldots$$

If you now look at the retained sequence

$$1\,0\,1\,1\,0\,1\,0\,1\,1\,0\ldots$$

you find that it is identical to the Golden Sequence.

We can do another magnification exercise on the Golden Sequence simply by underlining any pattern or subsequence. For example, suppose we choose "10" as our subsequence, and we underline it whenever it occurs in the Golden Sequence:

$$\underline{10}\,1\,\underline{10}\ \underline{10}\,1\,\underline{10}\,1\,\underline{10}\ \underline{10}\,1\,\underline{10}\ldots$$

If we now treat each 10 as a single symbol and we mark the number of places by which each pattern of 10 needs to be moved to overlap with the next 10, we get the sequence: 2122121 . . . (the first "10" needs to be moved two places to overlap with the second, the third is one place after the second, etc.). If we would now replace each 2 by a 1 and each 1 by a 0 in the new sequence, we recover the Golden Sequence. In other words, if we look at any pattern within the Golden Sequence, we discover that the same pattern is found in the sequence on another scale. Objects with this property, like the Russian Matrioshka dolls that fit one into the other, are known as *fractals*. The name "fractal" (from the Latin *fractus,* meaning "broken, fragmented") was coined by the famous Polish-French-American mathematician Benoit B. Mandelbrot, and it is a central concept in the geometry of nature and in the theory of highly irregular systems known as *chaos*.

Fractal geometry represents a brilliant attempt to describe the shapes and objects of the real world. When we look around us, very few forms can be described in terms of the simple figures of Euclidean geometry, such as straight lines, circles, cubes, and spheres. An old mathematical joke tells of a physicist who thought that he could become rich from betting at horse races by solving the exact equations of motion for the horses. After much work, he indeed managed to solve the equations—for spherical horses. Real horses, unfortunately, are not spherical, and neither are clouds, cauliflowers, or lungs. Similarly, lightning, rivers, and drainage systems do not travel in straight lines, and they all remind us of the branching of trees and of the human circulatory system. Examine, for example, the fantastically intricate branching of the "Dolmen in the Snow" (Figure 111), a painting by the German romantic painter Caspar David Friedrich (1774–1840; currently in the Gemäldegalerie Neue Meister in Dresden).

Figure 111

Mandelbrot's gigantic mental leap in formulating fractal geometry has been primarily in the fact that he recognized that all of these complex zigs and zags are not merely a nuisance but often the main mathematical characteristic of the morphology. Mandelbrot's first realization was the importance of *self-similarity*—the fact that many natural shapes display endless sequences of motifs repeating themselves within motifs, on many scales. The chambered nautilus (Figure 4) exhibits this property magnificently, as does a regular cauliflower—break off smaller and smaller pieces and, up to a point, they continue to look like the whole vegetable. Take a picture of a small piece of rock, and you will have a hard time recognizing that you are not looking at an entire mountain. Even the printed form of the continued fraction that is equal to the Golden Ratio has this property (Figure 112)—magnify the barely resolved symbols and you will see the same continued fraction. In all of these objects, zooming in does not smooth out the degree of roughness. Rather, the same irregularities characterize all scales.

At this point, Mandelbrot asked himself, how do you determine the dimensions of something that has such a fractal structure? In the world of Euclidean geometry, all the objects have dimensions that can be expressed as whole numbers. Points have zero dimensions, straight lines are one-dimensional, plane figures like triangles and pentagons are two-dimensional, and objects like spheres and

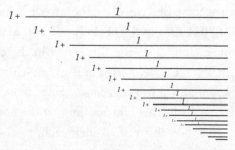

Figure 112

the Platonic solids are three-dimensional. Fractal curves like the path of a bolt of lightning, on the other hand, wiggle so aggressively that they fall somewhere between one and two dimensions. If the path is relatively smooth, then we can imagine that the fractal dimension would be close to one, but if it is very complex, then a dimension closer to two can be expected. These musings have turned into the by now-famous question: "How long is the coast of Britain?" Mandelbrot's surprising answer is that the length of the coastline actually depends on the length of your ruler. Suppose you start out with a satellite-generated map of Britain that is one foot on the side. You measure the length and convert it to the actual length by multiplying by the known scale of your map. Clearly this method will skip over any twists in the coastline that are too small to be revealed on the map. Equipped with a one-yard stick, you therefore start the long journey of actually walking along Britain's beaches, painstakingly measuring the length yard by yard. There is no doubt that the number you get now will be much larger than the previous one, since you managed to capture much smaller twists and turns. You immediately realize, however, that you would still be skipping over structures on smaller scales than one yard. The point is that every time you decrease the size of your ruler, you get a larger value for the length, because you always discover that there exists substructure on even smaller scale. This fact suggests that even the concept of length as representing size needs to be revisited when dealing with fractals. The contours of the coastline do not become a straight line upon magnification; rather, the crinkles persist on all scales and the length increases ad infinitum (or at least down to atomic scales).

This situation is exemplified beautifully by what could be thought of as the coastline of some imaginary land. The Koch snowflake is a curve first described by the Swedish mathematician Helge von Koch (1870–1924) in 1904 (Figure 113). Start with an equilateral triangle, one inch long on the side. Now in the middle of each side, construct a smaller triangle, with a side of one-third of an inch. This will give the Star of David in the second figure. Note that the original outline of the triangle was three inches long, while now it is composed of twelve segments, one-third of an inch each, so that the total length is now four inches. Repeat the same procedure consecutively—on each side of a triangle place a new one, with a side length that is one-third that of the previous one. Each time, the length of the outline increases by a factor of 4/3 to infinity, in spite of the fact that it borders a finite area. (We can show that the area converges to eight-fifths that of the original triangle.)

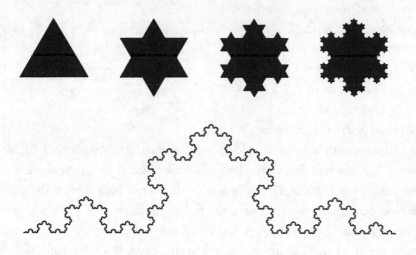

Figure 113

The realization of the existence of fractals raised the question of the dimensions that should be associated with them. The fractal dimension is really a measure of the wrinkliness of the fractal, or of how fast length, surface, or volume increases if we measure it with respect to ever-decreasing scales. For example, we feel intuitively that the Koch curve (bottom of Figure 113) takes up more space than a one-dimensional line but less space than the two-dimensional square. But how can it have an intermediate dimension? There is, after all, no whole number between 1 and 2. This is where Mandelbrot followed a concept first introduced in 1919 by the German mathematician Felix Hausdorff (1868–1942), a concept that at first appears mind boggling—fractional dimensions. In spite of the initial shock we may experience from such a notion, fractional dimensions were precisely the tool needed to characterize the degree of irregularity, or fractal complexity, of objects.

In order to obtain a meaningful definition of the self-similarity dimension or fractal dimension, it helps to use the familiar whole-number dimensions 0, 1, 2, 3 as guides. The idea is to examine how many small objects make up a larger object in any number of dimensions. For example, if we bisect a (one-dimensional) line, we obtain two segments (for a reduction factor of f = ½). When we divide a (two-dimensional) square into

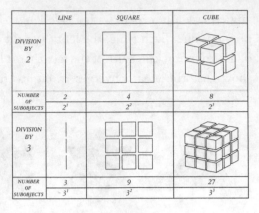

	LINE	SQUARE	CUBE
DIVISION BY 2			
NUMBER OF SUBOBJECTS	2 2^1	4 2^2	8 2^3
DIVISION BY 3			
NUMBER OF SUBOBJECTS	3 3^1	9 3^2	27 3^3

Figure 114

subsquares with half the side length (again a reduction factor f = ½), we get 4 = 2^2 squares. For a side length of one-third (f = ⅓), there are 9 = 3^2 subsquares (Figure 114). For a (three-dimensional) cube, a division into cubes of half the edge-length (f = ½) produces 8 = 2^3 cubes, and one-third the length (f = ⅓) produces 27 = 3^3 cubes (Figure 114). If you examine all of these examples, you find that there is a relation between the number of subobjects, n, the length reduction factor, f, and the dimension, D. The relation is simply $n = (1/f)^D$. (I give another form of this relation in Appendix 7.) Applying the same relation to the Koch snowflake gives a fractal dimension of about 1.2619. As it turns out, the coastline of Britain also has a fractal dimension of about 1.26. Fractals therefore serve as models for real coastlines. Indeed, pioneering chaos theorist Mitch Feigenbaum, of Rockefeller University in New York, exploited this fact to help produce in 1992 the revolutionary *Hammond Atlas of the World.* Using computers to do as much as possible unassisted, Feigenbaum examined fractal satellite data to determine which points along coastlines have the greatest significance. The result—a map of South America, for example, that is better than 98 percent accurate, compared to the more conventional 95 percent scored by older atlases.

For many fractals in nature, from trees to the growth of crystals, the main characteristic is branching. Let us examine a highly simplified model for this ubiquitous phenomenon. Start with a stem of unit length, which divides into two branches of length ½ at 120 degrees

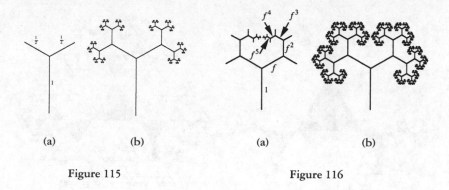

(a)   (b)   (a)   (b)

Figure 115   Figure 116

(Figure 115). Each branch further divides in a similar fashion, and the process goes on without bound.

If instead of a length reduction factor of ½ we had chosen a somewhat larger number (e.g., 0.6), the spaces among the different branches would have been reduced, and eventually branches would overlap. Clearly, for many systems (e.g., a drainage system or a blood circulatory system), we may be interested in finding out at what reduction factor precisely do the branches just touch and start to overlap, as in Figure 116. Surprisingly (or maybe not, by now), this happens for a reduction factor that is equal precisely to *one over the Golden Ratio,* $1/\phi = 0.618\ldots$ (A short proof is given in Appendix 8.) This is known as a *Golden Tree,* and its fractal dimension turns out to be about 1.4404. The Golden Tree and similar fractals composed of simple lines cannot be resolved very easily with the naked eye after several iterations. The problem can be partially resolved by using two-dimensional figures like *lunes* (Figure 117) instead of lines. At each step, you can use a copying machine equipped with an image reduction feature to produce lunes

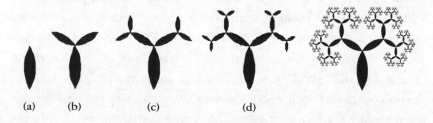

(a)   (b)   (c)   (d)

Figure 117   Figure 118

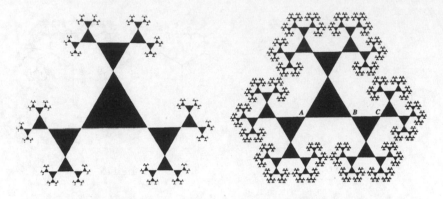

Figure 119   Figure 120

Figure 121   Figure 122

reduced by a factor $1/\phi$. The resulting image, a Golden Tree composed of lunes, is shown in Figure 118.

Fractals can be constructed not just from lines but also from simple planar figures such as triangles and squares. For example, you can start with an equilateral triangle with a side of unit length and at each corner attach a new triangle with a side length of ½. At each of the free corners of the second-generation triangles, attach a triangle with a side length of ¼, and so on (Figure 119). Again, you may wonder at what reduction factor do the three boughs start to touch, as in Figure 120, and again the answer turns out to be $1/\phi$. Precisely the same situation occurs if you build a similar fractal using a square (Figure 121)—overlapping occurs when the reduction factor is $1/\phi = 0.618\ldots$ (Figure 122).

Furthermore, all the unfilled white rectangles in the last figure are Golden Rectangles. We therefore find that while in Euclidean geometry the Golden Ratio originated from the pentagon, in fractal geometry it is associated even with simpler figures like squares and equilateral triangles.

Once you get used to the concept, you realize that the world around us is full of fractals. Objects as diverse as the profiles of the tops of forests on the horizon and the circulatory system in a kidney can be described in terms of fractal geometry. If a particular model of the universe as a whole known as eternal inflation is correct, then even the entire universe is characterized by a fractal pattern. Let me explain this concept very briefly, giving only the broad-brush picture. The inflationary theory, originally advanced by Alan Guth, suggests that when our universe was only a tiny fraction of a second old, an unbridled expansion stretched our region of space to a size that is actually much larger than the reach of our telescopes. The driving force behind this stupendous expansion is a very peculiar state of matter called a false vacuum. A ball on top of a flat hill, as in Figure 123, can symbolically describe the situation. For as long as the universe remained in the false vacuum state (the ball was on the hilltop), it expanded extremely rapidly, doubling in size every tiny fraction of a second. Only when the ball rolled down the hill and into the surrounding, lower-energy "ditch" (representing symbolically the fact that the false vacuum decayed) did the tremendous expansion stop. According to the inflationary model, what we call *our* universe was caught in the false vacuum state for a very brief period, during which it expanded at a

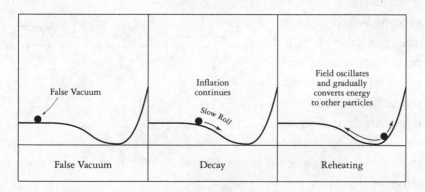

Figure 123

fantastic rate. Eventually the false vacuum decayed, and our universe resumed the much more leisurely expansion we observe today. All the energy and subatomic particles of our universe were generated during oscillations that followed the decay (represented schematically in the third drawing in Figure 123). However, the inflationary model also predicts that the rate of expansion while in the false vacuum state is much faster than the rate of decay. Consequently, the fate of a region of false vacuum can be illustrated schematically as in Figure 124. The universe started with some region of false vacuum. As time progressed, some part (a third in the figure) of the region has decayed to produce a "pocket universe" like our own. At the same time, the regions that stayed in the false vacuum state continued to expand, and by the time represented schematically by the second bar in Figure 124, each one of them was actually the size of the whole first bar. (This is not shown in the figure because of space constraints.) Moving in time from the second bar to the third, the central pocket universe continued to evolve slowly as in the standard big bang model of our universe. Each of the remaining two regions of false vacuum, however, evolved in precisely the same way as the original region of false vacuum—some part of them decayed, producing a pocket universe. Each region of false vacuum expanded to become the same size as the first bar (again, not shown in the figure because of space constraints). An infinite number of pocket universes thus were produced, and a fractal pattern was generated—the same sequence of false vacua and pocket universes is replicated on ever-decreasing scales. If this model truly represents the evolution of the universe as a whole, then our pocket universe is but one out of an infinite number of pocket universes that exist.

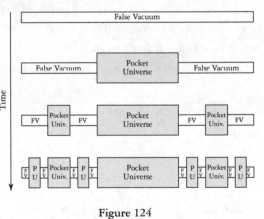

Figure 124

In 1990, North Carolina State University professor Jasper Memory published a poem entitled "Blake and Fractals" in the *Mathematics Magazine.* Referring to the mystic poet William Blake's line "To see a World in a Grain of Sand," Memory wrote:

> William Blake said he could see
> Vistas of infinity
> In the smallest speck of sand
> Held in the hollow of his hand.
> Models for this claim we've got
> In the work of Mandelbrot:
> Fractal diagrams partake
> Of the essence sensed by Blake.
> Basic forms will still prevail
> Independent of the Scale;
> Viewed from far or viewed from near
> Special signatures are clear.
> When you magnify a spot,
> What you had before, you've got.
> Smaller, smaller, smaller, yet,
> Still the same details are set;
> Finer than the finest hair
> Blake's infinity is there,
> Rich in structure all the way—
> Just as the mystic poets say.

Some of the modern applications of the Golden Ratio, Fibonacci numbers, and fractals reach into areas that are much more down to earth than the inflationary model of the universe. In fact, some say that the applications can reach even all the way into our pockets.

A GOLDEN TOUR OF WALL STREET

One of the best-known attempts to use the Fibonacci sequence and the Golden Ratio in the analysis of stock prices is associated with the name of Ralph Nelson Elliott (1871–1948). An accountant by profession, Elliott held various executive positions with railroad companies, primarily in Central America. A serious alimentary tract illness that left him bedridden forced him into retirement in 1929. To occupy his mind, Elliott started to analyze in great detail the rallies and plunges of the Dow Jones Industrial Average. During his lifetime, Elliott witnessed the roaring bull market of the 1920s followed by the Great Depression. His detailed analyses led him to conclude that market fluctuations were not random. In particular, he noted: "the stock market is a creation of man and therefore reflects human idiosyncrasy." Elliott's main observation was that, ultimately, stock market patterns reflect cycles of human optimism and pessimism.

On February 19, 1935, Elliott mailed a treatise entitled *The Wave Principle* to a stock market publication in Detroit. In it he claimed to have identified characteristics which "furnish a principle that determines the trend and gives clear warning of reversal." The treatise eventually developed into a book with the same title, which was published in 1938.

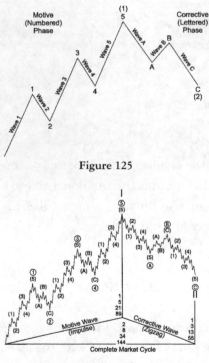

Figure 125

Figure 126

Elliott's basic idea was relatively simple. He claimed that market variations can be characterized by a fundamental pattern consisting of five waves during an upward ("optimistic") trend (marked by numbers in Figure 125) and three waves during a downward ("pessimistic") trend (marked by letters in Figure 125). Note that 5, 3, 8 (the total number of waves) are all Fibonacci numbers. Elliott further asserted that an examination of the fluctuation on shorter and shorter time scales reveals that the same pattern repeats itself (Figure 126), with all the numbers of the constituent wavelets corresponding to higher Fibonacci numbers. Identifying 144 as "the highest number of practical value," the breakdown of a complete market cycle, according to Elliott, might look as follows. A generally upward trend consisting of five major waves, twenty-one intermediate waves, and eighty-nine minor waves (Figure 126) is followed by a generally downward phase with three major, thirteen intermediate, and fifty-five minor waves (Figure 126).

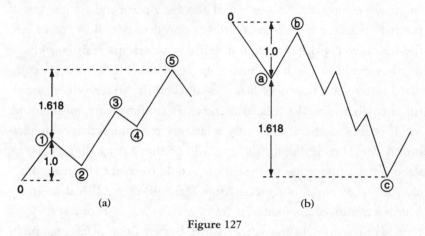

(a)   (b)

Figure 127

Some recent books that attempt to apply Elliott's general ideas to actual trading strategies go even further. They use the Golden Ratio to calculate the extreme points of maximum and minimum that can be expected (although not necessarily reached) in market prices at the end of upward or downward trends (Figure 127). Even more sophisticated algorithms include a logarithmic spiral plotted on top of the daily market fluctuations, in an attempt to represent a relationship between price and time. All of these forecasting efforts assume that the Fibonacci sequence and the Golden Ratio somehow provide the keys to the operation of mass psychology. However, this "wave" approach does suffer from some shortcomings. The Elliott "wave" usually is subjected to various (sometimes arbitrary) stretchings, squeezings, and other alterations by hand to make it "forecast" the real-world market. Investors know, however, that even with the application of all the bells and whistles of modern portfolio theory, which is supposed to maximize the returns for a decided-on level of risk, fortunes can be made or lost in a heartbeat.

You may have noticed that Elliott's wave interpretation has as one of its ingredients the concept that each part of the curve is a reduced-scale version of the whole, a concept central to fractal geometry. Indeed, in 1997, Benoit Mandelbrot published a book entitled *Fractals and Scaling in Finance: Discontinuity, Concentration, Risk,* which introduced well-defined fractal models into market economics. Mandelbrot built on the known fact that fluctuations in the stock market look the same when charts are enlarged or reduced to fit the same price and time scales. If you look at such a chart from a distance that does not allow you to read the scales, you will not be able to tell if it represents daily, weekly, or hourly variations. The main innovation in Mandelbrot's theory, as compared to standard portfolio theory, is in its ability to reproduce tumultuous trading as well as placid markets. Portfolio theory, on the other hand, is able to characterize only relatively tranquil activity. Mandelbrot never claimed that his theory could predict a price drop or rise on a specific day but rather that the model could be used to estimate probabilities of potential outcomes. After Mandelbrot published a simplified description of his model in *Scientific American* in February 1999, a myriad of responses from readers ensued. Robert Ihnot of Chicago probably expressed the bewilderment of many when he wrote: "If we know that a stock will go from $10 to $15 in a given amount of time, it doesn't matter how we interpose the fractals, or whether the graph looks authentic or not. The important thing is that we could buy at $10 and sell at $15. Everyone should now be rich, so why are they not?"

Elliott's original wave principle represented a bold if somewhat naïve attempt to identify a pattern in what appears otherwise to be a rather random process. More recently, however, Fibonacci numbers and randomness have had an even more intriguing encounter.

RABBITS AND COIN TOSSES

The defining property of the Fibonacci sequence—that each new number is the sum of the previous two numbers—was obtained from an unrealistic description of the breeding of rabbits. Nothing in this definition hinted that this imaginary rabbit sequence would find its way into so many natural and cultural phenomena. There was even less, however, to suggest that experimentation with the basic properties of the sequence themselves could provide a gateway to understanding the mathematics of disordered systems. Yet this was precisely what happened in 1999. Computer scientist Divakar Viswanath, then a postdoctoral fellow at the Mathematical Sciences Research Institute in Berkeley, California, was bold enough to ask a "what if?" question that led unexpectedly to the discovery of a new special number: 1.13198824. . . . The beauty of Viswanath's discovery lies primarily in the simplicity of its central idea. Viswanath merely asked himself: Suppose you start with the two numbers 1, 1, as in the original Fibonacci sequence, but now instead of adding the two numbers to get the third, you flip a coin to decide whether to add them or to subtract the last number from the previous one. You can decide, for example, that "heads" means to add (giving 2 as the third number) and "tails" means to subtract (giving 0 as the third number). You can continue with the same procedure, each time flipping a coin to decide whether to add or subtract the last number to get a new one. For example, the series of tosses HTTHHTHTTH will produce the sequence 1, 1, 2, –1, 3, 2, 5, –3, 2, –5, 7, 2. On the other hand, the (rather unlikely) series of tosses HHHHHHHHHHHH . . . will produce the original Fibonacci sequence.

In the Fibonacci sequence, terms increase rapidly, like a power of the Golden Ratio. Recall that we can calculate the seventeenth number in the sequence, for example, by raising the Golden Ratio to the seventeenth power, dividing by the square root of 5, and rounding off the result to the nearest whole number (which gives 1597). Since Viswanath's sequences were generated by a totally random series of coin tosses, however, it was not at all obvious that a smooth growth pattern would be obtained, even if we ignore the minus signs and take only the absolute value of the numbers. To his own surprise, however, Viswanath found that if he ignored the minus signs, the values of the numbers in his random sequences still increased in a clearly defined and predictable rate. Specifically, with essentially 100 percent probability, the one hundredth number in any of the sequences generated in this way was always close to the one hundredth power of the peculiar number 1.13198824 . . . , and the higher the term was in the sequence, the closer it came to the corresponding power of 1.13198824. . . . To actually calculate this strange number, Viswanath had to use fractals and to rely on a powerful mathematical theorem that was formulated in the early 1960s by mathematicians Hillel Furstenberg of the Hebrew University in Jerusalem and Harry Kesten of Cornell University. These two mathematicians proved that for an entire class of randomly generated sequences, the absolute value of a number high in the sequence gets closer and closer to the appropriate power of some fixed number. However, Furstenberg and Kesten did not know how to calculate this fixed number; Viswanath discovered how to do just that.

The importance of Viswanath's work lies not only in the discovery of a new mathematical constant, a significant feat in itself, but also in the fact that it illustrates beautifully how what appears to be an entirely random process can lead to a fully deterministic result. Problems of this type are encountered in a variety of natural phenomena and electronic devices. For example, stars like our own Sun produce their energy in nuclear "furnaces" at their centers. However, for us actually to see the stars shining, bundles of radiation, known as photons, have to make their way from the stellar depths to the surface. Photons do not simply fly through the star at the speed of light. Rather, they bounce around, being scattered and absorbed and reemitted by all the electrons and atoms of gas in their way, in a seemingly random fashion. Yet the net result is that after a random walk, which in the case of the Sun takes some 10 million years, the radiation escapes the star. The power emitted by the Sun's surface determined (and continues to determine) the temperature on Earth's surface and allowed life to emerge. Viswanath's work and the research on random Fibonaccis that followed provide additional tools for the mathematical machinery that explains disordered systems.

There is another important lesson to be learned from Viswanath's discovery—even an eight-hundred-year-old, seemingly trivial mathematical problem can still surprise you.

9

IS GOD A
MATHEMATICIAN?

I should attempt to treat human vice and folly geometrically . . .
the passions of hatred, anger, envy, and so on, considered in themselves,
follow from the necessity and efficacy of nature. . . . I shall, therefore,
treat the nature and strength of the emotion in exactly the same manner,
as though I were concerned with lines, planes and solids.
—BARUCH SPINOZA (1632–1677)

Two and two the mathematician continues to make four, in spite of the
whine of the amateur for three, or the cry of the critic for five.
—JAMES MCNEILL WHISTLER (1834–1903)

Euclid defined the Golden Ratio because he was interested in using this simple proportion for the construction of the pentagon and the pentagram. Had this remained the Golden Ratio's only application, the present book would have never been written. The delight we derive from this concept today is based primarily on the element of *surprise*. The Golden Ratio turned out to be, on one hand, the simplest of the continued fractions (but also the "most irrational" of all irrational numbers) and, on the other, the heart of an endless number of complex natural phenomena. Somehow the Golden Ratio always makes an unexpected appearance at the juxtaposition of the simple and the complex, at the intersection of Euclidean geometry and fractal geometry.

The sense of gratification provided by the Golden Ratio's surprising emergences probably comes as close as we could expect to the sensuous visual pleasure we obtain from a work of art. This fact raises the question of what type of aesthetic judgment can be applied to mathematics or, even more specifically, what did the famous British mathematician Godfrey Harold Hardy (1877–1947) actually mean when he said: "The mathematician's patterns, like the painter's or the poet's, must be beautiful."

This is not an easy question. When I discussed the psychological experiments that tested the visual appeal of the Golden Rectangle, I deliberately avoided the term "beautiful." I will adopt the same strategy here, because of the ambiguity associated with the definition of beauty. The extent to which beauty is in the eye of the beholder when referring to mathematics is exemplified magnificently by a story presented in the excellent 1981 book *The Mathematical Experience* by Philip J. Davis and Reuben Hersh.

In 1976, a delegation of distinguished mathematicians from the United States was invited to the People's Republic of China for a series of talks and informal meetings with Chinese mathematicians. The delegation subsequently issued a report entitled "Pure and Applied Mathematics in the People's Republic of China." By "pure," mathematicians usually refer to the type of mathematics that at least on the face of it has absolutely no direct relevance to the world outside the mind. At the same time, we should realize that Penrose tilings and random Fibonaccis, for example, provide two of the numerous examples of "pure" mathematics turning into "applied." One of the dialogues in the delegation's report, between Princeton mathematician Joseph J. Kohn and one of his Chinese hosts, is particularly illuminating. The dialogue was on the topic of the "beauty of mathematics," and it took place at the Shanghai Hua-Tung University.

> *Kohn:* Should you not present beauty of mathematics? Couldn't it inspire students? Is there room for the beauty of science?

Answer: The first demand is production.

Kohn: That is no answer.

Answer: Geometry was developed for practice. The evolution
of geometry could not satisfy science and
technology; in the seventeenth century, Descartes
discovered analytical geometry. He analyzed pistons
and lathes and also the principles of analytical
geometry. Newton's work came out of the
development of industry. Newton said, "The basis of
any theory is social practice." There is no theory of
beauty that people agree on. Some people think one
thing is beautiful, some another. Socialist
construction is a beautiful thing and stimulates
people here. Before the Cultural Revolution some of
us believed in the beauty of mathematics but failed
to solve practical problems; now we deal with water
and gas pipes, cables and rolling mills. We do it for
the country and the workers appreciate it. It is a
beautiful feeling.

Since, as this dialogue starkly indicates, there is hardly any formal, accepted description of aesthetic judgment in mathematics and how it should be applied, I prefer to discuss only one particular element in mathematics that invariably gives pleasure to nonexperts and experts alike—the element of surprise.

MATHEMATICS SHOULD SURPRISE

In a letter written on February 27, 1818, the English Romantic poet John Keats (1795–1821) wrote: "Poetry should surprise by a fine excess and not by Singularity—it should strike the Reader as a wording of his own highest thoughts, and appear almost a Remembrance." Unlike poetry, however, mathematics more often tends to delight when it exhibits an unanticipated result rather than when conforming to the reader's own expectations. In addition, the pleasure derived from mathematics is related in many cases to the surprise felt upon perception of totally unexpected relationships and unities. A mathematical relation known as Benford's law provides a wonderful case study for how all of these elements combine to produce a great sense of satisfaction.

Take a look, for example, in the *World Almanac,* at the table of "U.S. Farm Marketings by State" for 1999. There is a column for "Crops" and one for "Livestock and Products." The numbers are given in U.S. dollars. You would have thought that the numbers from 1 to 9 should occur with the same frequency among the first digits of all the listed marketings. Specifically, the numbers starting with 1 should constitute about one-ninth of all the listed numbers, as would numbers starting with 9. Yet, if you count them, you will find that the number 1 appears as the first digit in 32 percent of the numbers (instead of the expected 11 percent if all digits occurred equally often). The number 2 also appears more frequently than its fair share—appearing in 19 percent of the numbers. The number 9, on the other hand, appears only in 5 percent of the numbers—less than expected. You may think that finding this result in one table is surprising, but hardly shocking, until you examine a few more pages in the *Almanac* (the numbers above were taken from the 2001 edition). For example, if you look at the table of the death toll of "Some Major Earthquakes," you will find that the numbers starting with 1 constitute about 38 percent of all the numbers, and those starting with 2 are 18 percent. If you choose a totally different table, such as the one for the population in Massachusetts in places of 5,000 or more, the numbers start with 1 about 36 percent of the time and with 2 about 16.5 percent of the time. At the other end, in all of these tables the number 9 appears first only in about 5 percent of the numbers, far less than the expected 11 percent. How is it possible that tables describing such diverse and apparently random data all have the property that the number 1 appears as the first digit 30-some percent of the time and the number 2 around 18 percent of the time? The situation becomes even more puzzling when you examine still larger databases. For example, accounting professor Mark Nigrini of the Cox School of Business at Southern Methodist University, Dallas, examined the populations of 3,141 counties in the 1990 U.S. Census. He found that the number 1 appeared as the first digit in about 32 percent of the numbers, 2 appeared in about 17 percent, 3 in 14 percent, and 9 in less than 5 percent. Analyst Eduardo Ley of Resources for the Future in Washington, D.C., found very similar numbers for the Dow Jones Industrial Average in the years 1990 to 1993. And if all of this is not dumfounding enough, here is another amazing fact. If you examine the list of, say, the first two thousand Fibonacci numbers, you will find that the number 1 appears as the first digit 30 percent of the time, the number 2 appears 17.65 percent, 3 appears 12.5 percent, and the values continue to decrease, with 9 appearing 4.6 percent of the time as first digit. In fact, Fibonacci numbers are more likely to start with 1, with the other numbers decreasing in popularity *in precisely the same manner as the just-described random selections of numbers!*

Astronomer and mathematician Simon Newcomb (1835–1909) first discovered this "first-digit phenomenon" in 1881. He noticed that books of logarithms in the library, which were used for calculations, were considerably dirtier at the beginning (where numbers starting with 1 and 2 were printed) and progressively cleaner throughout. While this might be expected with bad novels abandoned by bored readers, in the case of mathematical tables they simply indicated a more frequent appearance of numbers starting with 1 and 2. Newcomb, however, went much further than merely noting this fact; he came up with an actual *formula* that was supposed to give the probability that a random number begins with a particular digit. That formula (presented in Appendix 9) gives for 1 a probability of 30 percent; for 2, about 17.6 percent; for 3, about 12.5 percent; for 4, about 9.7 percent; for 5, about 8 percent; for 6, about 6.7 percent; for 7, about 5.8 percent; for 8, about 5 percent; and for 9, about 4.6 percent. Newcomb's 1881 article in the *American Journal of Mathematics* and the "law" he discovered went entirely unnoticed, until fifty-seven years later, when physicist Frank Benford of General Electric rediscovered the law (apparently independently) and tested it with extensive data on river basin areas, baseball statistics, and even numbers appearing in *Reader's Digest* articles. All the data fit the postulated formula amazingly well, and hence this formula is now known as Benford's law.

Not all lists of numbers obey Benford's law. Numbers in telephone books, for example, tend to begin with the same few digits in any given region. Even tables of square roots of numbers do not obey the law. On the other hand, chances are that if you collect all the numbers appearing on the front pages of several of your local newspapers for a week, you will obtain a pretty good fit. But why should it be this way? What do the populations of towns in Massachusetts have to do with death tolls from earthquakes around the globe or with numbers appearing in the *Reader's Digest*? Why do the Fibonacci numbers also obey the same law?

Attempts to put Benford's law on a firm mathematical basis have proven to be much more difficult than expected. One of the key obstacles has been precisely the fact that not all lists of numbers obey the law (even the preceding examples from the *Almanac* do not obey the law precisely). In his *Scientific American* article describing the law in 1969, University of Rochester mathematician Ralph A. Raimi concluded that "the answer remains obscure."

The explanation finally emerged in 1995–1996, in the work of Georgia Institute of Technology mathematician Ted Hill. Hill became first interested in Benford's law while preparing a talk on surprises in probability in the early 1990s. When describing to me his experience, Hill said: "I started working on this problem as a recreational experiment, but a few people warned me to be careful, because Benford's law can become addictive." After a few years of work it finally dawned on him that rather than looking at numbers from one given source, the *mixture* of data was the key. Hill formulated the law statistically, in a new form: "If distributions are selected at random (in any unbiased way) and random samples are taken from each of these distributions, then the significant-digit frequencies of the *combined sample* will converge to Benford's distribution, even if some of the individual distributions selected do not follow the law." In other words, suppose you assemble random collections of numbers from a hodgepodge of distributions, such as a table of square roots, a table of the death toll in notable aircraft disasters, the populations of counties, and a table of air distances between selected world cities. Some of these distributions do not obey Benford's law by themselves. What Hill proved, however, is that as you collect ever more of such numbers, the digits of these numbers will yield frequencies that conform ever closer to the law's predictions. Now, why do Fibonacci numbers also follow Benford's law? After all, they are fully determined by a recursive relation and are not random samples from random distributions.

Well, in this case it turns out that this conformity with Benford's law is not a unique property of the Fibonacci numbers. If you examine a large number of powers of 2 ($2^1 = 2$, $2^2 = 4$, $2^3 = 8$, etc.), you'll see that they also obey Benford's law. This should not be so surprising, given that the Fibonacci numbers themselves are obtained as powers of the Golden Ratio (recall that the nth Fibonacci number is close to $\phi^n/\sqrt{5}$). In fact, we can prove that sequences defined by a large class of recursive relations follow Benford's law.

Benford's law provides yet another fascinating example of pure mathematics transformed into applied. One interesting application is in the detection of fraud or fabrication of data in accounting and tax evasion. In a broad range of financial documents, data conform very closely to Benford's law. Fabricated data, on the other hand, very rarely do. Hill demonstrates how such fraud detection works with another simple example, using probability theory. In the first day of class in his course on probability, he asks students to do an experiment. If their mother's maiden name begins with A through L, they are to flip a coin 200 times and record the results. The rest of the class is asked to fake a sequence of 200 heads and tails. Hill collects the results the following day, and within a short time he is able to separate the genuine from the fake with 95 percent accuracy. How does he do that? Any sequence of 200 genuine coin tosses contains a run of six consecutive heads or six consecutive tails with a very high probability. On the other hand, people trying to fake a sequence of coin tosses very rarely believe that they should record such a sequence.

A recent case in which Benford's law was used to uncover fraud involved an American leisure and travel company. The company's audit director discovered something that looked odd in claims made by the supervisor of the company's healthcare department. The first two digits of the healthcare payments showed a suspicious spike in numbers starting with 65 when checked for conformity to Benford's law. (A more detailed version of the law predicts also the frequency of the second and higher digits; see Appendix 9.) A careful audit revealed thirteen fraudulent checks for amounts between $6,500 and $6,599. The District

Attorney's office in Brooklyn, New York, also used tests based on Benford's law to detect accounting fraud in seven New York companies.

Benford's law contains precisely some of the ingredients of surprise that most mathematicians find attractive. It reflects a simple but astonishing fact—that the distribution of first digits is extremely peculiar. In addition, that fact turned out to be difficult to explain. Numbers, with the Golden Ratio as an outstanding example, sometimes provide a more instantaneous gratification. For example, many professional and amateur mathematicians are fascinated by primes. Why are primes so important? Because the "Fundamental Theorem of Arithmetic" states that every whole number larger than 1 can be expressed as a product of prime numbers. (Note that 1 is not considered a prime.) For example, $28 = 2 \times 2 \times 7$; $66 = 2 \times 3 \times 11$; and so on. Primes are so rooted in the human comprehension of mathematics that in his book *Cosmos,* when Carl Sagan (1934–1996) had to describe what type of signal an intelligent civilization would transmit into space he chose as an example the sequence of primes. Sagan wrote: "It is extremely unlikely that any natural physical process could transmit radio messages containing prime numbers only. If we received such a message we would deduce a civilization out there that was at least fond of prime numbers." The great Euclid proved more than two thousand years ago that infinitely many primes exist. (The elegant proof is presented in Appendix 10.) Yet most people will agree that some primes are more attractive than others. Some mathematicians, such as the French François Le Lionnais and the American Chris Caldwell, maintain lists of "remarkable" or "titanic" numbers. Here are just a few intriguing examples from the great treasury of primes:

- The number 1,234,567,891, which cycles through all the digits, is a prime.
- The 230th largest prime, which has 6,400 digits, is composed of 6,399 9s and only one 8.
- The number composed of 317 iterations of the digit 1 is a prime.
- The 713th largest prime can be written as $(10^{1951}) \times (10^{1975} + 1991991991991991991991991) + 1$, and it was discovered in— you guessed it—1991.

From the perspective of this book, the connection between primes and Fibonacci numbers is of special interest. With the exception of the number 3, every Fibonacci number that is a prime also has a prime subscript (its order in the sequence). For example, the Fibonacci number 233 is a prime, and it is the thirteenth (also a prime) number in the sequence. The converse, however, is not true: The fact that the subscript is a prime does not necessarily mean that the number is also a prime. For cxample, the nineteenth number (19 is a prime) is 4181, and 4181 is not a prime—it is equal to 113×37.

The number of known Fibonacci primes has increased steadily over the years. In 1979, the largest known Fibonacci prime was the 531st in the sequence. By the mid-1990s, the largest known was the $2,971^{st}$; and in 2001, the $81,839^{th}$ number was shown to be a prime with 17,103 digits. So, is there an infinite number of Fibonacci primes (as there is an infinite number of primes, in general)? No one actually knows, and this is probably the greatest unsolved mathematical mystery about Fibonacci numbers.

THE UNREASONABLE POWER OF MATHEMATICS

The collection of dialogues *Intentions* contains the aesthetic philosophy of the famous playwright and poet Oscar Wilde (1854–1900). In that collection, the dialogue "The Decay of Lying" is a particularly provocative presentation of Wilde's ideas on "the new aesthetics." In the conclusion of this dialogue, one of the characters (Vivian) summarizes:

> Life imitates Art far more than Art imitates Life. This results not merely from Life's imitative instinct, but from the fact that the selfconscious aim of Life is to find expression, and that Art offers it certain beautiful forms through which it may realize that energy. It is a theory that has never been put forward before, but it is extremely fruitful, and throws an entirely new light upon the history of Art.
>
> It follows, as a corollary from this, that external Nature also imitates Art. The only effects that she can show us are effects that

we have already seen through poetry, or in paintings. This is the secret of Nature's charm, as well as the explanation of Nature's weakness.

We could almost substitute "Mathematics" for "Art" in this passage and obtain a statement that reflects the reality with which many outstanding minds have struggled. Mathematics appears at first glance to be just too effective. In Einstein's own words: "How is it possible that mathematics, a product of human thought that is independent of experience, fits so excellently the objects of physical reality?" Another outstanding physicist, Eugene Wigner (1902–1995), known for his many contributions to nuclear physics, delivered in 1960 a famous lecture entitled "The Unreasonable Effectiveness of Mathematics in the Physical Sciences." We have to wonder, for example, how is it possible that planets in their orbits around the Sun were found to follow a curve (an ellipse) that had been explored by the Greek geometers long before Kepler's laws were discovered? Why does the explanation of the existence of quasi-crystals rely on the Golden Ratio, a concept conceived by Euclid for purely mathematical purposes? Is it not astounding that the structure of so many galaxies containing billions of stars follows closely Bernoulli's favorite curve—the magnificent logarithmic spiral? And the most astonishing of all: Why are the laws of physics themselves expressible as mathematical equations in the first place?

But this is not all. Mathematician John Forbes Nash (now world famous as the subject of the book and film biography *A Beautiful Mind*), for example, shared the 1994 Nobel Prize in economics because his mathematical dissertation (written at age twenty-one!) outlining his "Nash Equilibrium" for strategic noncooperative games inaugurated a revolution in fields as diverse as economics, evolutionary biology, and political science. What is it that makes mathematics work so well?

The recognition of the extraordinary "effectiveness" of mathematics even made it into a hysterically funny passage in Samuel Beckett's novel *Molloy,* about which I have a personal story. In 1980, two colleagues from the University of Florida and I wrote a paper about neutron stars, which are extremely compact and dense astronomical objects that result from the gravitational collapse of the cores of massive stars. The paper was more mathematical than the garden variety of astronomical papers, and consequently we decided to add an appropriate motto on the first page. The motto read:

Extraordinary how mathematics help you . . .
—SAMUEL BECKETT, *Molloy*

The line was cited as being taken from the first of the trilogy of novels *Molloy, Malone Dies,* and *The Unnamable* by the famous writer and playwright Samuel Beckett (1906–1989). All three novels, incidentally, represent a search for self—a hunt for identity by writers through writing. We are led to observe the characters in states of decay while they pursue a meaning for their existence.

Papers in astrophysics very rarely have mottoes. Consequently, we received a letter from the editor of *The Astrophysical Journal* informing us that while he liked Beckett, too, he did not quite see the necessity of including the motto. We replied that we would leave the decision of whether to publish the motto or not entirely to him, and the paper eventually was published with the motto in the December 15 issue. Here, however, is the full passage from *Molloy:*

And in winter, under my greatcoat, I wrapped myself in swathes of newspaper, and did not shed them until the earth awoke, for good, in April. The Times Literary Supplement was admirably adapted to this purpose, of a neverfailing toughness and impermeability. Even farts made no impression on it. I can't help it, gas escapes from my fundament on the least pretext, it's hard not to mention it now and then, however great my distaste. One day I counted them. Three hundred and fifteen farts in nineteen hours, or an average of over sixteen farts an hour. After all it's not excessive. Four farts every fifteen minutes. It's nothing. Not even one fart every four minutes. It's unbelievable. Damn it, I hardly fart at all, I should never have mentioned it. Extraordinary how mathematics help you to know yourself.

The history of mathematics has produced at least two attempts, philosophically very different, to answer the question of the incredible power of mathematics. The answers are also related to the fundamental issue of the actual *nature* of mathematics. A comprehensive discussion of these topics can fill entire volumes and is certainly beyond the scope of this book. I will therefore give only a brief description of some of the main lines of thought and present my personal opinion.

One view on the nature of mathematics, traditionally dubbed the "Platonic view," is that mathematics is universal and timeless, and its existence is an objective fact, independent of us humans. According to this Platonic view, mathematics has always been out there in some abstract world, for humans to simply discover, just as Michelangelo thought that his sculptures existed inside the marble and he merely uncovered them. The Golden Ratio, Fibonacci numbers, Euclidean geometry, and Einstein's equations are all a part of this Platonic reality that transcends the human mind. Supporters of this Platonic view regard the famous Austrian logician Kurt Gödel (1906–1978) also as a wholehearted Platonist. They point out that not only did he say about mathematical concepts that "they, too, may represent an aspect of objective reality" but that his "incompleteness theorems" by themselves could be taken as arguments in favor of the Platonic view. These theorems, probably the most celebrated results in the whole of logic, show that for any formal axiomatic system (e.g., number theory) there exist statements formulable in its language that *it cannot either prove or disprove.* In other words, number theory, for example, is "incomplete" in the sense that there are *true* statements of number theory that the theory's methods of proof are incapable of demonstrating. To prove them we must jump to a higher and richer system, in which again other true statements can be made that cannot be proved, and so on ad infinitum. Computer scientist and author Douglas R. Hofstadter phrased this succinctly in his fantastic book *Gödel, Escher, Bach: An Eternal Golden Braid:* "Provability is a weaker notion than truth." In this sense, there will never be a formal method of determining for every mathematical proposition whether it is absolutely true, any more than there is a way to determine whether a theory in physics is absolutely true. Oxford's mathematical physicist Roger Penrose is among those who believe that

Gödel's theorems argue powerfully for the very existence of a Platonic mathematical world. In his wonderfully thought-provoking book *Shadows of the Mind* Penrose says: "Mathematical truth is not determined arbitrarily by the rules of some 'man-made' formal system, but has an absolute nature, and lies beyond any such system of specifiable rules." To which he adds that: "Support for the Platonic viewpoint . . . was an important part of Gödel's initial motivations." Twentieth-century British mathematician G. H. Hardy also believed that the human function is to "discover or observe" mathematics rather than to invent it. In other words, the abstract landscape of mathematics was there, waiting for mathematical explorers to reveal it.

One of the proposed solutions to the mystery of the effectiveness of mathematics in explaining nature relies on an intriguing modification of the Platonic ideas. This "modified Platonic view" argues that the laws of physics are expressed as mathematical equations, the structure of the universe is a fractal, galaxies arrange themselves in logarithmic spirals, and so on, because mathematics is the *universe's language.* Specifically, mathematical objects are still assumed to exist objectively, quite independent of our knowledge of them, but instead of placing mathematics entirely in some mythical abstract plane, at least some parts of it would be placed in the real cosmos. If we want to communicate with intelligent civilizations 10,000 light-years away, all we have to do is transmit the number 1.6180339887 . . . and be sure that they will understand what we mean, because the universe has undoubtedly imposed the same mathematics on them. God is indeed a mathematician.

This modified Platonic view was precisely the belief expressed by Kepler (colored by his religious inclinations), when he wrote that geometry "supplied God with patterns for the creation of the world, and passed over to Man along with the image of God; and was not in fact taken in through the eyes." Galileo Galilei had similar thoughts:

> Philosophy is written in this grand book—I mean the universe—which stands continually open to our gaze, but it cannot be understood unless one first learns to comprehend the language and interpret the characters in which it is written. It is written in the language of mathematics, and its characters are triangles, circles,

and other geometrical figures, without which it is humanly impossible to understand a single word of it; without these, one is wandering about in a dark labyrinth.

The mystic poet and artist William Blake had a rather different opinion of this mathematician God. Blake utterly despised scientific explanations of nature. To him, Newton and the scientists who followed him merely conspired to unweave the rainbow, to conquer all mysteries of human life by rules. Accordingly, in Blake's powerful etching "The Ancient of Days" (Figure 128; currently at the Pierpont Morgan Library, New York), he depicts an evil God who wields a compass not to establish universal order but rather to clip the wings of imagination.

Figure 128

Kepler and Galileo, however, were definitely not the last mathematicians to adopt this "modified" version of the Platonic view, nor were such views limited to those who, like Newton, took for granted the existence of a Divine Mind. The great French mathematician, astronomer, and physicist Pierre-Simon de Laplace (1749–1827) wrote in his *Théorie Analitique des Probabilités* (Analytic theory of probabilities; 1812):

> Given for one instant an intelligence which comprehends all the forces by which nature is animated and the respective positions of the beings which compose it, if moreover this intelligence were vast

enough to submit these data to analysis, it would embrace in the same formula both the movements of the largest bodies in the universe and those of the lightest atom.

This was the same Laplace who replied to Napoleon Bonaparte: "Sire, I have no need for that hypothesis," when the emperor remarked that there is no mention of the creator in Laplace's large book on celestial mechanics.

Very recently, IBM mathematician and author Clifford A. Pickover wrote in his lively book *The Loom of God:* "I do not know if God is a mathematician, but mathematics is the loom upon which God weaves the fabric of the universe. . . . The fact that reality can be described or approximated by *simple* mathematical expressions suggests to me that nature has mathematics at its core."

Supporters of the "modified Platonic view" of mathematics like to point out that, over the centuries, mathematicians have produced (or "discovered") numerous objects of pure mathematics with absolutely no application in mind. Decades later, these mathematical constructs and models were found to provide solutions to problems in physics. Penrose tilings and non-Euclidean geometries are beautiful testimonies to this process of mathematics unexpectedly feeding into physics, but there are many more.

There were also many cases of feedback between physics and mathematics, where a physical phenomenon inspired a mathematical model that later proved to be the explanation of an entirely different physical phenomenon. An excellent example is provided by the phenomenon known as Brownian motion. In 1827, British botanist Robert Brown (1773–1858) observed that when pollen particles are suspended in water, they get into a state of agitated motion. This effect was explained by Einstein in 1905 as resulting from the collisions that the colloidal particles experience with the molecules of the surrounding fluid. Each single collision has a negligible effect, because the pollen grains are millions of times more massive than the water molecules, but the persistent bombardment has a cumulative effect. Amazingly, the same model was found to apply to the motions of stars in star clusters. There the Brownian motion is produced by the cumulative effect of many stars passing by any given star, with each passage altering the motion (through gravitational interaction) by a tiny amount.

There exists, however, an entirely different view (from that of the modified Platonic view) on the nature of mathematics and the reason for its effectiveness. According to this view (which is intricately related to dogmas labeled "formalism" and "constructivism" in the philosophy of mathematics), mathematics has no existence outside the human brain. Mathematics as we know it is nothing but a human invention, and an intelligent civilization elsewhere in the universe might have developed a radically different construct. Mathematical objects have no objective reality—they are imaginary. In the words of the great German philosopher Immanuel Kant: "The ultimate truth of mathematics lies in the possibility that its concepts can be constructed by the human mind." In other words, Kant emphasizes the *freedom* aspect of mathematics, the freedom to postulate and to invent patterns and structures.

This view of mathematics as a human invention has become popular in particular with modern psychologists. For example, French researcher and author Stanislas Dehaene concludes in his interesting 1997 book *The Number Sense* that "intuitionism [which to him is synonymous with mathematics as a human invention] seems to me to provide the best account of the relations between arithmetic and the human brain." Similarly, the last sentence in the book *Where Mathematics Comes From* (2000) by the University of California, Berkeley, linguist George Lakoff and psychologist Rafael E. Núñez reads: "The portrait of mathematics has a human face." These conclusions are based primarily on the results of psychological experiments and on neurological studies of the functionality of the brain. Experiments show that babies have innate mechanisms for recognizing numbers in small sets and that children acquire simple arithmetical capabilities spontaneously, even without much formal instruction. Additionally, the inferior parietal cortex has been identified as the area of the brain that hosts the neural circuitry involved in symbolic numerical capabilities. This area of both cerebral hemispheres is located anatomically at the junction of neural connections from touch, vision, and audition. In patients suffering from a rare form of seizure while performing arithmetic manipulations (known as epilepsia arithmetices), brain wave measurements (electroencephalograms) show abnormalities in the inferior parietal cortex. Similarly, lesions in this region affect mathematical ability, writing, and spatial coordination.

Even if based on physiology and psychology, the view of mathematics as a human invention of no intrinsic reality still needs to answer the two intriguing questions: Why is mathematics so powerful in explaining the universe, and how is it possible that even some of the purest products of mathematics are found eventually to fit physical phenomena like a glove?

The "human inventionist" reply to both of these questions is also based on a biological model: evolution and natural selection. The idea here is that progress in understanding the universe and the formulation of mathematical laws that describe phenomena within it have been achieved via an extended and tortuous evolutionary process. Our current model of the universe is the result of a long evolution that involved many false starts and blind alleys. Natural selection has weeded out mathematical models that did not fit the observations and experiments and left only the successful ones. According to this view, all "theories" of the universe are in fact nothing but "models" whose attributes are determined solely by their success in fitting the observational and experimental data. Kepler's crazy model of the solar system in *Mysterium Cosmographicum* was acceptable, as long as it could explain and predict the behavior of the planets.

The success of pure mathematics turned into applied mathematics, in this picture, merely reflects an overproduction of concepts, from which physics has selected the most adequate for its needs—a true survival of the fittest. After all, "inventionists" would point out, Godfrey H. Hardy was always proud of having "never done anything 'useful.'" This opinion of mathematics is apparently espoused also by Marilyn vos Savant, the "world record holder" in IQ—an incredible 228. She is quoted as having said "I'm beginning to think simply that mathematics can be invented to describe anything, and matter is no exception."

In my humble opinion, neither the modified Platonic view nor the natural selection view provides a fully satisfactory answer (at least in the way both are traditionally formulated) to the mystery of the effectiveness of mathematics.

To claim that mathematics is purely a human invention and is successful in explaining nature *only* because of evolution and natural selection ignores some important facts in the nature of mathematics and in the history of theoretical models of the universe. First, while the mathematical rules (e.g., the axioms of geometry or of set theory) are indeed creations of the human mind, once those rules are specified, we lose our freedom. The definition of the Golden Ratio emerged originally from the axioms of Euclidean geometry; the definition of the Fibonacci sequence from the axioms of the theory of numbers. Yet the fact that the ratio of successive Fibonacci numbers converges to the Golden Ratio was *imposed* on us—humans had no choice in the matter. Therefore, mathematical objects, albeit imaginary, do have *real* properties. Second, the explanation of the unreasonable power of mathematics cannot be based entirely on evolution in the restricted sense. For example, when Newton proposed his theory of gravitation, the data that he was trying to explain were at best accurate to three significant figures. Yet his mathematical model for the force between any two masses in the universe achieved the incredible precision of better than one part in a million. Hence, that particular model was not *forced* on Newton by existing measurements of the motions of planets, nor did Newton force a natural phenomenon into a preexisting mathematical pattern. Furthermore, natural selection in the common interpretation of that concept does not quite apply either, because it was not the case that five competing theories were proposed, of which one eventually won. Rather, Newton's was the only game in town!

The modified Platonic view, on the other hand, faces different types of challenges.

First, there is the important conceptual issue that the modified Platonic view does not really offer any *explanation* to the power of mathematics. The question is simply transformed into a belief in the mathematical underpinning of the physical world. Mathematics is simply *assumed* to be the symbolic counterpart of the universe. Roger Penrose, who as I noted before is himself a powerful supporter of the Platonic world of mathematical forms, agrees that the "puzzling precise underlying role that the Platonic mathematical world has in the physical world" remains a mystery. Oxford University physicist David Deutsch turns the question somewhat around. In his insightful 1997 book *The Fabric of Reality,* he wonders: "in a reality composed of physics and understood by the methods of science, where does mathematical certainty come from?" Penrose adds to the effectiveness of mathematics two more mysteries. In his book *Shadows of the Mind,* he wonders: "How it is that perceiving beings can arise from out of the physical world," and "how it is that mentality is able seemingly to 'create' mathematical concepts out of some kind of mental model." These intriguing questions, which are entirely outside the scope of the present book, deal with the origin of consciousness and the perplexing ability of our rather primitive mental tools to gain access into the Platonic world (which to Penrose is an objective reality).

The second problem encountered by the modified Platonic view is related to the question of *universality.* To what extent are we certain that the laws that the universe must obey have to be presented by mathematical equations of the type we have formulated? Until very recently, probably most physicists on the face of the Earth would have argued that history has shown that equations are the *only* way in which the laws of physics can be expressed. This situation may change, however, with the impending publication of the book *A New Kind of Science* by Stephen Wolfram. Wolfram, one of the most innovative thinkers in scientific computing and in the theory of complex systems, has been best known for the development of Mathematica, a computer program/system that allows a range of calculations not accessible before. After ten years of virtual silence, Wolfram is about to emerge with a provocative book that makes the bold claim that he can replace the basic infrastructure of science. In a world used to more than three hundred years of science being dominated by mathematical equations as the basic building blocks of models for nature, Wolfram proposes simple computer programs instead. He suggests that nature's main secret is the use of simple programs to generate complexity.

Wolfram's book was not out yet at the time of this writing, but from a long conversation I had with him and from an interview he gave to science writer Marcus Chown, I can safely conclude that his work has many far-reaching implications. From the restricted point of view of its reflection on Platonism, however, Wolfram's work points out that at the very least, the particular mathematical world that many thought exists out there, and which was believed to underlie physical reality, may not be unique. In other words, there definitely can exist descriptions of nature that are very different from the one we have. Mathematics as we know it captures only a tiny part of the vast space of all possible simple sets of rules that might describe the workings of the cosmos.

If both the modified Platonic view and the natural selection interpretation have difficulties in attempting to explain the striking effectiveness of mathematics, is there an exposition that works?

I believe that the explanation has to rely on concepts borrowed from both points of view rather than on adopting one or the other. The situation here is very similar to the historical attempts in physics to explain the nature of light. The lesson from this piece of history of science is so profound that I will describe it now briefly.

Newton's first paper was on optics, and he continued to work on this subject for most of his life. In 1704 he published the first edition of his book *Opticks,* which he later revised three times. Newton proposed a "particle theory of light," in which light was assumed to be made up of tiny, hard particles, that obey the same laws of motion as do billiard balls. In Newton's words: "Even the rays of light seem to be hard bodies." Two famous experiments at the beginning of the twentieth century discovered the photoelectric effect and the Compton effect, and provided strong support for the idea of particles of light. The photoelectric effect is a process in which electrons in a piece of metal absorb sufficient energy from light to allow them to escape. Einstein's explanation of this effect in 1905 (which won him the 1922 Nobel Prize for Physics) showed that light delivers the energy to the electrons in a grainy fashion, in indivisible units of energy. Thus, the *photon*—the particle of light—was introduced. Physicist Arthur Holly Compton (1892–1962) analyzed in 1918 to 1925 the scattering of X rays from electrons both experimentally and theoretically. His work (which won him the 1927 Nobel Prize for Physics) further confirmed the existence of the photon.

But there was another theory of light—a *wave theory*—in which light was assumed to behave like waves of water in a pond. This theory was most strongly advocated by the Dutch physicist Christiann Huygens (1629–1695). The wave theory did not have much going for it until the physicist and physician Thomas Young (1773–1829) discovered *interference* in 1801. The phenomenon itself is quite simple. Suppose you dip the index fingers of both hands periodically into the water in a pond. Each finger will create a sequence of concentric ripples; crest and trough will follow each other in the form of expanding rings. At points where a crest emanating from one finger intersects a crest from the other, you get the two waves to enhance each other ("constructive interference"). At points where a crest overlaps with a trough, they annihilate each other ("destructive interference"). A detailed analysis of the fixed pattern that emerges shows that along the central line (between the two fingers), there is constructive interference. To either side, lines of destructive interference alternate with lines of constructive interference.

In the case of light, destructive interference simply means dark lines. Young, a child prodigy who spoke eleven languages by age sixteen, performed an experiment in which he passed light through two slits and demonstrated that the light on the viewing surface was "divided by dark stripes."

Young's results, which were followed by impressive theoretical work by French engineer Augustin Fresnel in 1815 to 1820, initiated a conversion of physicists to the wave theory. Later experiments conducted by the French physicist Léon Foucault in 1850 and by American physicist Albert Michelson in 1883 showed unambiguously that the refraction of light as it passes from air to water also behaves precisely as predicted by the wave theory. More important, the Scottish physicist James Clerk Maxwell (1831–1879) published in 1864 a comprehensive theory of electromagnetism that predicted the existence of propagating electromagnetic waves moving at the speed of light. He went on to propose that light itself is an electromagnetic wave. Finally, between 1886 and 1888, the German physicist Heinrich Herz proved experimentally that light was indeed the electromagnetic wave predicted by Maxwell.

So, what is light? Is it a pure bombardment by particles (photons) or a pure wave? Really, it is neither. Light is a more complicated physical phenomenon than any single one of these concepts, which are based on classical physical models, can describe. To describe the propagation of light and to understand phenomena like interference, we can and have to use the electromagnetic wave theory. When we want to discuss the interaction of light with elementary particles, however, we have to use the photon description. This picture, in which the particle and wave descriptions complement each other, has become known as the *wave-particle duality.* The modern quantum theory of light has unified the classical notions of waves and particles in the concept of probabilities. The electromagnetic field is represented by a wave function, which gives the probabilities of finding the field in certain modes. The photon is the energy associated with these modes.

Returning now to the question of the nature of mathematics and the reason for its effectiveness, I believe that the same type of complementarity should be applied. Mathematics was *invented,* in the sense that the "rules of the game" (the sets of axioms) are man-made. But once invented, it took on a life of its own, and humans had (and still have) to discover all of its properties, in the spirit of the Platonic view. The endless list of unexpected appearances of the Golden Ratio, the numberless mathematical relations obeyed by the Fibonacci numbers, and the fact that we still do not know if there are infinitely many Fibonacci primes provide ample evidence for this discovery quest.

Wolfram holds very similar views. I asked him specifically whether he thought mathematics was "invented" or "discovered." He replied: "If there wasn't much choice in selecting this particular set of rules then it would have made sense to say that it was discovered, but since there was much choice, and our mathematics is merely historically based, I have to say that it was invented." The phrase "historically based" in this context is crucial since it implies that the system of axioms on which our mathematics is based is the one that happened to emerge because of the arithmetic and geometry of the ancient Babylonians. This raises two immediate questions: (1) Why did the Babylonians develop these particular disciplines and not other sets of rules? And a rephrasing of the question on the effectiveness of mathematics: (2) Why were these disciplines and their offspring found to be useful at all for physics?

Interestingly, the answers to both of these questions may be related. Mathematics itself could have originated from a subjective human perception of how nature works. Geometry may simply reflect the human ability to easily recognize lines, edges, and curves. Arithmetic may represent the human aptitude to resolve discrete objects. In this picture, the mathematics that we have is a feature *of the biological details of humans and of how they perceive the cosmos.* Mathematics thus is, in some sense, the language of the universe—of the universe discerned by humans. Other intelligent civilizations out there might have developed totally different sets of rules, if their mechanisms of perception are very different from ours. For example, when one drop of water is added to another drop or one molecular cloud in the galaxy coalesces with another, they make only one drop or one cloud, not two. Therefore, if a civilization that is somehow fluid based exists, for it, one plus one does not necessarily equal two. Such a civilization may recognize neither the prime numbers nor the Golden Ratio. To give another example, there is hardly any doubt that had even just the gravity of Earth been much stronger than it actually is, the Babylonians and Euclid might have proposed a different geometry than the Euclidean. Einstein's theory of general relativity has taught us that in a much stronger gravitational field, space around us would be curved, not flat—light rays would travel along curved paths rather than along straight lines. Euclid's geometry emerged from his simple observations in Earth's weak gravity. (Other geometries, on curved surfaces, were formulated in the nineteenth century.)

Evolution and natural selection definitely played a cardinal role in our theories of the universe. This is precisely why we don't continue to adhere today to the physics of Aristotle. This is not to say, however, that the evolution was always continuous and smooth. The biological evolution of life on Earth was neither. Life's pathway was occasionally shaped by chance occurrences like mass extinctions. Impacts of astronomical bodies (comets or asteroids) several miles in diameter caused the dinosaurs to perish and paved the way for the dominance of the mammals. The evolution of theories of the universe was also sporadically punctuated by quantum leaps in understanding. Newton's theory of gravitation and Einstein's General Relativity ("I still can't see how he thought of it," said the late physicist Richard Feynman) are two perfect examples of such spectacular advances. How can we explain these miraculous achievements? The truth is that we can't. That is, no more than we can explain how, in a world of chess that was used to victories by margins of half a point or so, in 1971 Bobby Fischer suddenly demolished both chess grandmasters Mark Taimanov and Bent Larsen by scores of six points to nothing on his way to the world championship. We may find it equally difficult to comprehend how naturalists Charles Darwin (1809–1882) and Alfred Russel Wallace (1823–1913) independently had the inspiration to introduce the concept of evolution itself—the idea of a descent of all life from a common ancestral origin. We must simply recognize the fact that certain individuals are head and shoulders above the rest in terms of insight. Can, however, dramatic breakthroughs like Newton's and Einstein's be accommodated at all in a scenario of evolution and natural selection? They can, but in a somewhat less common interpretation of natural selection. While it is true that Newton's theory of universal gravitation had no contending theories to compete with at the time, it would not have survived to the present day had it not been the "fittest." Kepler, by contrast, proposed a very short-lived model for the Sun-planet interaction, in which the Sun spins on its axis flinging rays of magnetic power. These rays were supposed to grab on the planets and push them in a circle.

When these generalized definitions of evolution (allowing for quantum jumps) and natural selection (operating over extended periods of time) are adopted, I believe that the "unreasonable" effectiveness of mathematics finds an explanation. Our mathematics is the symbolic counterpart of the *universe we perceive,* and its power has been continuously enhanced by human exploration.

Jef Raskin, the creator of the Macintosh computer at Apple, emphasizes a different aspect—the evolution of human logic. In a 1998 essay on the effectiveness of mathematics, he concludes that "Human *logic* [emphasis added] was forced on us by the physical world and is therefore consistent with it. Mathematics derives from logic. This is why mathematics is consistent with the physical world."

In the play *Tamburlaine the Great,* a tale about a Machiavellian hero-villain who is at the same time sensitive and a vicious murderer, the great English playwright Christopher Marlowe (1564–1593) recognizes this human aspiration for understanding the cosmos:

Nature that framed us of four elements,
Warring within our breasts for regiment,
Doth teach us all to have aspiring minds:
Our souls, whose faculties can comprehend
The wondrous Architecture of the world:
And measure every wandering planet's course,
Still climbing after knowledge infinite,
And always moving as the restless spheres . . .

The Golden Ratio is a product of humanly invented geometry. Humans had no idea, however, into what magical fairyland this product was going to lead them. If geometry had not been invented at all, then we might have never known about the Golden Ratio. But then, who knows? It might have emerged as the output of a short computer program.

APPENDIX 1

We want to show that for any whole numbers p and q, such that p is larger than q, the three numbers: $p^2 - q^2$; $2pq$; $p^2 + q^2$ form a Pythagorean triple. In other words, we need to show that the sum of the squares of the first two is equal to the square of the third. For this we use the general identities that hold for any a and b:

$$(a + b)^2 = (a + b)(a + b) = a^2 + ab + ba + b^2 = a^2 + 2ab + b^2$$
$$(a - b)^2 = (a - b)(a - b) = a^2 - ab - ba + b^2 = a^2 - 2ab + b^2.$$

Based on these identities, the square of the first number is:

$$(p^2 - q^2)^2 = p^4 - 2p^2q^2 + q^4$$

and the sum of the first two squares is:

$$p^4 - 2p^2q^2 + q^4 + 4p^2q^2 = p^4 + 2p^2q^2 + q^4.$$

The square of the last number is:

$$(p^2 + q^2)^2 = p^4 + 2p^2q^2 + q^4.$$

We therefore see that the square of the third number is indeed equal to the sum of the squares of the first two, irrespective of the values of p and q.

APPENDIX 2

We want to prove that the diagonal and the side of the pentagon are incommensurable—they do not have any common measure.

The proof is by the general method of reductio ad absurdum described at the end of Chapter 2.

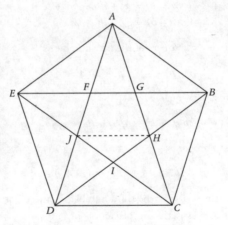

Let us denote the side of the pentagon $ABCDE$ by s_1 and the diagonal by d_1. From the properties of isosceles triangles you can easily prove that $AB = AH$ and $HC = HJ$. Let us now denote the side of the smaller pentagon $FGHIJ$ by s_2 and its diagonal by d_2. Clearly

$$AC = AH + HC = AB + HJ.$$

Therefore:

$$d_1 = s_1 + d_2 \text{ or } d_1 - s_1 = d_2.$$

If d_1 and s_1 have a common measure, it means that both d_1 and s_1 are some integer multiple of that common measure. Consequently, this is also a common measure of $d_1 - s_1$ and therefore of d_2. Similarly, the equalities

$$AG = HC = HJ$$
$$AH = AB$$

and

$$AH = AG + GH$$
$$AB = HJ + GH$$

give us

$$s_1 = d_2 + s_2$$

or

$$s_1 - d_2 = s_2.$$

Since based on our assumption the common measure of s_1 and d_1 is also a common measure of d_2, the last equality shows that it is also a common measure of s_2. We therefore find that the same unit that measures s_1 and d_1 also measures s_2 and d_2. This process can be continued ad infinitum, for smaller and smaller pentagons. We would obtain that the same unit that was a common measure for the side and diagonal of the *first* pentagon is also a common measure of all the other pentagons, irrespective of how tiny they become. Since this clearly cannot be true, it means that our initial assumption that the side and diagonal have a common measure was false—this completes the proof that s_1 and d_1 are incommensurable.

APPENDIX 3

The area of a triangle is half the product of the base and the height to that base. In the triangle *TBC* the base, *BC*, is equal to 2*a* and the height, *TA*, is equal to *s*. Therefore, the area of the triangle is equal to *s* × *a*. We want to show that if the square of the *pyramid's* height, h^2, is equal to the area of its triangular face, *s* × *a*, then *s*/*a* is equal to the Golden Ratio.

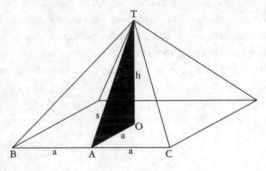

We have that

$$h^2 = s \times a.$$

Using the Pythagorean theorem in the right angle triangle *TOA*, we have

$$s^2 = h^2 + a^2.$$

We can now substitute for h^2 from the first equation to obtain

$$s^2 = s \times a + a^2.$$

Dividing both sides by a^2, we get:

$$(s/a)^2 = (s/a) + 1.$$

In other words, if we denote *s*/*a* by *x*, we have the quadratic equation:

$$x^2 = x + 1.$$

In Chapter 4 I show that this is precisely the equation defining the Golden Ratio.

APPENDIX 4

One of the theorems in *The Elements* demonstrates that when two triangles have the same angles, they are *similar*. Namely, the two triangles have precisely the same shape, with all their sides being proportional to each other. If one side of one triangle is twice as long as the respective side of the other triangle, then so are other sides. The two triangles *ADB* and *DBC* are similar (because they have the same angles). Therefore, the ratio *AB/DB* (ratio of the *sides* of the two triangles *ADB* and *DBC*) is equal to *DB/BC* (ratio of the *bases* of the same two triangles):

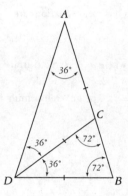

$$AB/DB = DB/BC.$$

But the two triangles are also isosceles, so that

$$DB = DC = AC.$$

We therefore find from the above two equalities that

$$AC/BC = AB/AC,$$

which means (according to Euclid's definition) that point *C* divides line *AB* in a Golden Ratio. Since *AD* = *AB* and *DB* = *AC*, we also have *AD/DB* = ϕ.

APPENDIX 5

Quadratic equations are equations of the form

$$ax^2 + bx + c = 0,$$

where a, b, c are arbitrary numbers. For example, in the equation $2x^2 + 3x + 1 = 0$, $a = 2, b = 3, c = 1$.

The general formula for the two solutions of the equation is

$$x_1 = \frac{-b + \sqrt{b^2 - 4ac}}{2a}$$

$$x_2 = \frac{-b - \sqrt{b^2 - 4ac}}{2a}.$$

In the above example

$$x_1 = \frac{-3 + \sqrt{9 - 8}}{4} = \frac{-2}{4} = -\frac{1}{2}$$

$$x_2 = \frac{-3 - \sqrt{9 - 8}}{4} = \frac{-4}{4} = -1.$$

In the equation we obtained for the Golden Ratio,

$$x^2 - x - 1 = 0,$$

we have $a = 1, b = -1, c = -1$. The two solutions therefore are:

$$x_1 = \frac{1 + \sqrt{1 + 4}}{2} = \frac{1 + \sqrt{5}}{2}$$

$$x_2 = \frac{1 - \sqrt{1 + 4}}{2} = \frac{1 - \sqrt{5}}{2}.$$

APPENDIX 6

The problem of the inheritance can be solved as follows. Let us denote the entire estate by E and the share (in bezants) of each son by x. (They all shared the inheritance equally.)

The first son received:

$$x = 1 + \frac{1}{7} (E - 1) .$$

The second son received:

$$x = 2 + \frac{1}{7} (E - 2 - x) .$$

Equating the two shares:

$$1 + \frac{1}{7} (E - 1) = 2 + \frac{1}{7} (E - 2 - x)$$

$$1 + \frac{E}{7} - \frac{1}{7} = 2 + \frac{E}{7} - \frac{2}{7} - \frac{x}{7}$$

and arranging:

$$\frac{x}{7} = \frac{6}{7}$$

$$x = 6.$$

Therefore, each son received 6 bezants.

Substituting in the first equation we have:

$$6 = 1 + \frac{1}{7} (E - 1)$$

$$6 = 1 + \frac{E}{7} - \frac{1}{7}$$

$$\frac{E}{7} = \frac{36}{7}$$

$$E = 36.$$

The total estate was 36 bezants. The number of sons was therefore $36/6 = 6$. Fibonacci's solution reads as follows:

The total inheritance has to be a number such that when 1 times 6 is added to it, it will be divisible by 1 plus 6, or 7; when 2 times 6 is added to it, it is divisible by 2 plus 6, or 8; when 3 times 6 is added, it is divisible by 3 plus 6, or 9, and so forth. The number is 36. $\frac{1}{7}$ of 36 minus $\frac{1}{7}$ is $35\frac{1}{7}$; plus 1 is $42\frac{1}{7}$, or 6; and this is the amount each son received; the total inheritance divided by the share of each son equals the number of sons, or $36/6$ equals 6.

APPENDIX 7

The relation between the number of subobjects, n, the length reduction factor, f, and the dimension, D, is

$$n = \left(\frac{1}{f}\right)^{D}.$$

If a positive number A is written as $A = 10^{L}$, then we call L the *logarithm* (base 10) of A, and we write it as log A. In other words, the two equations $A = 10^{L}$ and $L = \log A$ are entirely equivalent to each other. The rules of logarithms are:

(i) The logarithm of a *product* is the *sum* of the logarithms

$$\log (A \cdot B) = \log A + \log B.$$

(ii) The logarithm of a *ratio* is the *difference* of the logarithms

$$\log\left(\frac{A}{B}\right) = \log A - \log B.$$

(iii) The logarithm of a *power of a number* is the power *times* the logarithm of the number

$$\log A^{m} = m \log A.$$

Since $10^{0} = 1$, we have from the definition of the logarithm that log 1 = 0. Since $10^{1} = 10$, $10^{2} = 100$, and so on, we have that log 10 = 1, log 100 = 2, and so on. Consequently, the logarithm of any number between 1 and 10 is a number between 0 and 1; the logarithm of any number between 10 and 100 is a number between 1 and 2; and so on.

If we take the logarithm (base 10) of both sides in the above equation (describing the relation between n, f, and D), we obtain

$$\log n = D \log (1/f) = -D \log f.$$

Therefore, dividing both sides by log f

$$D = -\frac{\log n}{\log f}.$$

In the case of the Koch snowflake, for example, each curve contains four "subcurves" that are one-third in size; therefore $n = 4, f = \frac{1}{3}$ and we obtain

$$D = -\frac{\log 4}{\log\left(\frac{1}{3}\right)} = \frac{\log 4}{\log 3} = 1.2618595 \ldots.$$

APPENDIX 8

If we examine Figure 116(a), we see that the condition for the two branches to touch amounts to the simple requirement that the sum of all the *horizontal* lengths of the ever-decreasing branches with lengths starting with f^3 would be equal to the horizontal component of the large branch of length f. All the horizontal components are given by the total length multiplied by the cosine of 30 degrees. We therefore obtain:

$$f \cos 30° = f^3 \cos 30° + f^4 \cos 30° + f^5 \cos 30° + f^6 \cos 30° + \ldots$$

Dividing by cos 30° we obtain

$$f = f^3 + f^4 + f^5 + f^6 + \ldots$$

The sum on the right-hand side is the sum of an infinite *geometric* series (each term is equal to the previous term multiplied by a constant factor) in which the first term is f^3, and the ratio of two consecutive terms is f. In general, the sum S of an infinite geometric sequence in which the first term is a, and the ratio of consecutive terms q, is equal to

$$S = \frac{a}{1-q} .$$

For example, the sum of the sequence

$$1 + \frac{1}{2} + \frac{1}{4} + \frac{1}{8} + \frac{1}{16} + \cdots$$

in which $a = 1$ and $q = \frac{1}{2}$ is equal to

$$S = \frac{1}{1 - 1/2} = \frac{1}{1/2} = 2 .$$

In our case we find from the equation above:

$$f = \frac{f^3}{1-f} .$$

Dividing both sides by f, we get

$$1 = \frac{f^2}{1-f} \ .$$

Multiplying by $(1-f)$ and arranging, we obtain the quadratic equation:

$$f^2 + f - 1 = 0,$$

with the positive solution

$$\frac{\sqrt{5}-1}{2} \ ,$$

which is $1/\phi$.

APPENDIX 9

Benford's law states that the probability P that digit D appears in the *first* place is given by (logarithm base 10):

$$P = \log (1 + 1/D).$$

Therefore, for $D = 1$

$$P = \log (1 + 1) = \log 2 = 0.30.$$

For $D = 2$

$$P = \log (1 + \tfrac{1}{2}) = \log 1.5 = 0.176,$$

And so on. For $D = 9$,

$$P = \log (1 + \tfrac{1}{9}) = \log (\tfrac{10}{9}) = 0.046.$$

The more general law says, for example, that the probability that the first three digits are 1, 5, and 8 is:

$$P = \log (1 + \tfrac{1}{158}) = 0.0027.$$

APPENDIX 10

Euclid's proof that infinitely many primes exist is based on the method of reductio ad absurdum. He began by assuming the contradictory—that only a finite number of primes exist. If that is true, however, then one of them must be the largest prime. Let us denote that prime by P. Euclid then constructed a new number by the following process: He multiplied together all the primes from 2 up to (and including) P, and then he added 1 to the product. The new number is therefore

$$2 \times 3 \times 5 \times 7 \times 11 \ldots \times P + 1.$$

By the original assumption, this number must be composite (not a prime), because it is obviously larger than P, which was assumed to be the largest prime. Consequently, this number must be divisible by at least one of the existing primes. However, from its construction, we see that if we divide this number by any of the primes up to P, this will leave a remainder 1. The implication is, that if the number is indeed composite, some prime larger than P must divide it. However, this conclusion contradicts the assumption that P is the largest prime, thus completing the proof that there are infinitely many primes.

FURTHER READING

*It is only shallow people who do not judge by
appearances. The mystery of the world is the visible,
not the invisible.*

—OSCAR WILDE (1854–1900)

Most of the books and articles selected here are from the nontechnical literature. The few that are more technical were chosen on the basis of some special features. I have also listed a few websites that contain interesting material.

1: PRELUDE TO A NUMBER

Ackermann, F. "The Golden Section," *Mathematical Monthly*, 2 (1895): 260–264.

Dunlap, R. A. *The Golden Ratio and Fibonacci Numbers.* Singapore: World Scientific, 1997.

Fowler, D. H. "A Generalization of the Golden Section," *The Fibonacci Quarterly*, 20 (1982): 146–158.

Gardner, M. *The Second Scientific American Book of Mathematical Puzzles & Diversions.* Chicago: University of Chicago Press, 1987.

Ghyka, M. *The Geometry of Art and Life.* New York: Dover Publications, 1977.

Grattan-Guinness, I. *The Norton History of the Mathematical Sciences.* New York: W. W. Norton & Company, 1997.

Herz-Fischler, R. *A Mathematical History of the Golden Number.* Mineola, NY: Dover Publications, 1998.

Hoffer, W. "A Magic Ratio Occurs Throughout Art and Nature," *Smithsonian* (December 1975): 110–120.

Hoggatt, V. E., Jr. "Number Theory: The Fibonacci Sequence," in *Yearbook of Science and the Future.* Chicago: *Encyclopaedia Britannica*, 1977, 178–191.

Huntley, H. E. *The Divine Proportion.* New York: Dover Publications, 1970.

Knott, R. *www.mcs.surrey.ac.uk/Personal/R.Knott/Fibonacci/fib.html.*

Knott, R. *www.mcs.surrey.ac.uk/Personal/R.Knott/Fibonacci/fibnet2.html.*

Markowski, G. "Misconceptions about the Golden Ratio," *College Mathematics Journal,* 23 (1992): 2–19.

Ohm, M. *Die reine Elementar-Mathematik.* Berlin: Jonas Veilags-Buchhandlung, 1835.

Runion, G. E. *The Golden Section.* Palo Alto: Dale Seymour Publications, 1990.

2: THE PITCH AND THE PENTAGRAM

library.thinkquest.org/27890/mainIndex.html.

search.britannica.com/search?query=fibonacci.

Barrow, J. D. *Pi in the Sky.* Boston: Little, Brown and Company, 1992.

Beckmann, P. *A History of* π. Boulder, CO: Golem Press, 1977.

Boulger, W. "Pythagoras Meets Fibonacci," *Mathematics Teacher,* 82 (1989): 277–282.

Boyer, C. B. *A History of Mathematics.* New York: John Wiley & Sons, 1991.

Burkert, W. *Lore and Science in Ancient Pythagoreanism.* Cambridge, MA: Harvard University Press, 1972.

Conway, J. H., and Guy, R. K. *The Book of Numbers.* New York: Copernicus, 1996.

Dantzig, T. *Number: The Language of Science.* New York: The Free Press, 1954.

de la Füye, A. *Le Pentagramme Pythagoricien, Sa Diffusion, Son Emploi dans le Syllaboire Cuneiform.* Paris: Geuthner, 1934.

Guthrie, K. S. *The Pythagorean Sourcebook and Library.* Grand Rapids, MI: Phanes Press, 1988.

Ifrah, G. *The Universal History of Numbers.* New York: John Wiley & Sons, 2000.

Maor, E. *e: The Story of a Number.* Princeton, NJ: Princeton University Press, 1994.

Paulos, J. A. *Innumeracy.* New York: Vintage Books, 1988.

Pickover, C. A. *Wonders of Numbers.* Oxford: Oxford University Press, 2001.

Schimmel, A. *The Mystery of Numbers.* Oxford: Oxford University Press, 1994.

Schmandt-Besserat, D. "The Earliest Precursor of Writing," *Scientific American* (June 1978): 38–47.

Schmandt-Besserat, D. "Reckoning Before Writing," *Archaeology,* 32–33 (1979): 22–31.

Singh, S. *Fermat's Enigma.* New York: Anchor Books, 1997.

Stanley, T. *Pythagoras.* Los Angeles: The Philosophical Research Society, 1970.

Strohmeier, J., and Westbrook, P. *Divine Harmony.* Berkeley, CA: Berkeley Hills Books, 1999.

Turnbull, H. W. *The Great Mathematicians.* New York: Barnes & Noble, 1993.

von Fritz, K. "The Discovery of Incommensurability of Hipposus of Metapontum," *Annals of Mathematics,* 46 (1945): 242–264.

Wells, D. *Curious and Interesting Numbers.* London: Penguin Books, 1986.

Wells, D. *Curious and Interesting Mathematics.* London: Penguin Books, 1997.

3: UNDER A STAR-Y-POINTING PYRAMID?

Beard, R. S. "The Fibonacci Drawing Board Design of the Great Pyramid of Gizeh," *The Fibonacci Quarterly,* 6 (1968): 85–87.

Burton, D. M. *The History of Mathematics: An Introduction.* Boston: Allyn and Bacon, 1985.

Doczi, O. *The Power of Limits.* Boston: Shambhala, 1981.

Fischler, R. "Théories Mathématiques de la Grande Pyramide," *Crux Mathematicorum,* 4 (1978): 122–129.

Fischler, R. "What Did Herodotus Really Say? or How to Build (a Theory of) the Great Pyramid," *Environment and Planning B,* 6 (1979): 89–93.

Gardner, M. *Fads and Fallacies in the Name of Science.* New York: Dover Publications, 1957.

Gazalé, M. J. *Gnomon.* Princeton, NJ: Princeton University Press, 1999.

Gillings, R. J. *Mathematics in the Time of the Pharaohs.* New York: Dover Publications, 1972.

Goff, B. *Symbols of Prehistoric Mesopotamia.* New Haven, CT: Yale University Press, 1963.

Hedian, H. "The Golden Section and the Artist," *The Fibonacci Quarterly,* 14 (1976): 406–418.

Lawlor, R. *Sacred Geometry.* London: Thames and Hudson, 1982.

Mendelssohn, K. *The Riddle of the Pyramids.* New York: Praeger Publishers, 1974.

Petrie, W. *The Pyramids and Temples of Gizeh.* London: Field and Tuer, 1883.

Piazzi Smyth, C. *The Great Pyramid.* New York: Gramercy Books, 1978.

Schneider, M. S. *A Beginner's Guide to Constructing the Universe.* New York: Harper Perennial, 1995.

Spence, K. "Ancient Egyptian Chronology and the Astronomical Orientation of the Pyramids," *Nature,* 408 (2000): 320–324.

Stewart, I. "Counting the Pyramid Builders," *Scientific American* (September 1998): 98–100.

Verheyen, H. F. "The Icosahedral Design of the Great Pyramid," in *Fivefold Symmetry.* Singapore: World Scientific, 1992, 333–360.

Wier, S. K. "Insights from Geometry and Physics into the Construction of Egyptian Old Kingdom Pyramids," *Cambridge Archaeological Journal,* 6 (1996): 150–163.

4: THE SECOND TREASURE

Borissavlievitch, M. *The Golden Number and the Scientific Aesthetics of Architecture.* London: Alec Tiranti, 1958.

Bruckman, P. S. "Constantly Mean," *The Fibonacci Quarterly,* 15 (1977): 236.

Coxeter, H. S. M. *Introduction to Geometry.* New York: John Wiley & Sons, 1963.

Cromwell, P. R. *Polyhedra.* Cambridge: Cambridge University Press, 1997.

Dixon, K. *Mathographics.* New York: Dover Publications, 1987.

Ghyka, M. *L'Esthetique des proportions dans la nature et dans les arts.* Paris: Gallimard, 1927.

Heath, T. *A History of Greek Mathematics.* New York: Dover Publications, 1981.

Heath, T. *The Thirteen Books of Euclid's Elements.* New York: Dover Publications, 1956.

Jowett, B. *The Dialogues of Plato.* Oxford: Oxford University Press, 1953.

Kraut, R. *The Cambridge Companion to Plato.* Cambridge: Cambridge University Press, 1992.

Lasserre, F. *The Birth of Mathematics in the Age of Plato.* London: Hutchinson, 1964.

Pappas, T. *The Joy of Mathematics.* San Carlos, CA: Wide World Publishing, 1989.

Trachtenberg, M., and Hyman, I. *Architecture: From Prehistory to Post Modernism/The Western Tradition.* New York: Harry N. Abrams, 1986.

Zeising, A. *Der goldne Schnitt.* Halle: Druck von E. Blochmann & Son in Dresden, 1884.

5: SON OF GOOD NATURE

cedar.evansville.edu/~ck6/index.html.

Adler, I., Barabe, D., and Jean, R. V. "A History of the Study of Phyllotaxis," *Annals of Botany,* 80 (1997): 231–244.

Basin, S. L. "The Fibonacci Sequence as It Appears in Nature," *The Fibonacci Quarterly,* 1 (1963): 53–64.

Brousseau, Brother A. *An Introduction to Fibonacci Discovery.* Aurora, SD: The Fibonacci Association, 1965.

Bruckman, P. S. "Constantly Mean," *The Fibonacci Quarterly,* 15 (1977): 236.

Coxeter, H. S. M. "The Golden Section, Phyllotaxis, and Wythoff's Game," *Scripta Mathematica,* 19 (1953): 135–143.

Coxeter, H. S. M. *Introduction to Geometry.* New York: John Wiley & Sons, 1963.

Cook, T. A. *The Curves of Life.* New York: Dover Publications, 1979.

Devlin, K. *Mathematics.* New York: Columbia University Press, 1999.

Douady, S., and Couder, Y. "Phyllotaxis as a Physical Self-Organized Process," *Physical Review Letters,* 68 (1992): 2098–2101.

Dunlap, R. A. *The Golden Ratio and Fibonacci Numbers.* Singapore: World Scientific, 1997.

Fibonacci, L. P. *The Book of Squares.* Orlando, FL: Academic Press, 1987.

"The Fibonacci Numbers," *Time,* April 4, 1969, 49–50.

Gardner, M. *Mathematical Circus.* New York: Alfred A. Knopf, 1979.

Gardner, M. "The Multiple Fascination of the Fibonacci Sequence," *Scientific American* (March 1969): 116–120.

Garland, T. H. *Fascinating Fibonaccis.* White Plains, NY: Dale Seymour Publications, 1987.

Gies, J., and Gies, F. *Leonard of Pisa and the New Mathematics of the Middle Ages.* New York: Thomas Y. Crowell Company, 1969.

Hoggatt, V. E. Jr. "Number Theory: The Fibonacci Sequence," Chicago: *Encyclopaedia Britannica,* Yearbook of Science and the Future, 1977, 178–191.

Hoggatt, V. E. Jr., and Bicknell-Johnson, M. "Reflections Across Two and Three Glass Plates," *The Fibonacci Quarterly,* 17 (1979): 118–142.

Horadam, A. F. "Eight Hundred Years Young," *The Australian Mathematics Teacher,* 31 (1975): 123–134.

Jean, R. V. *Mathematical Approach to Pattern and Form in Plant Growth.* New York: John Wiley & Sons, 1984.

O'Connor, J. J. and Robertson, E. F. *www-history.mcs.st-andrews.ac.uk/history/Mathematicians/Fibonacci.html.*

Pickover, C. A. *Keys to Infinity.* New York: John Wiley & Sons, 1995.

Rivier, N., Occelli, R., Pantaloni, J., and Lissowdki, A. "Structure of Binard Convection Cells, Phyllotaxis and Crystallography in Cylindrical Symmetry," *Journal Physique,* 45 (1984): 49–63.

Singh, P. "The So-Called Fibonacci Numbers in Ancient and Medieval India," *Historia Mathematica,* 12 (1985): 229–244.

Smith, D. E. *History of Mathematics.* New York: Dover Publications, 1958.

Stewart, I. "Fibonacci Forgeries," *Scientific American* (May 1995): 102–105.

Stewart, I. *Life's Other Secret.* New York: John Wiley & Sons, 1998.

Thompson, D. W. *On Growth and Form.* New York: Dover Publications, 1992.

Vajda, S. *Fibonacci & Lucas Numbers, and the Golden Section.* Chichester: Ellis Horwood Limited, 1989.

Vorob'ev, N. N. *Fibonacci Numbers.* New York: Blaisdell, 1961.

6: THE DIVINE PROPORTION

Arasse, D. *Leonardo Da Vinci.* New York: Konecky & Konecky, 1998.

Beer, A. and Beer, P., eds., *Kepler: Four Hundred Years.* Vistas in Astronomy, vol. 18, New York: Pergamon Press, 1975.

Calvesi, M. *Piero Della Francesca.* New York: Rizzoli, 1998.

Caspar, M. *Kepler.* New York: Dover Publications, 1993.

Cromwell, P. R. *Polyhedra.* Cambridge: Cambridge University Press, 1997.

Gingerich, O. "Kepler, Galilei, and the Harmony of the World," in *Music and Science in the Age of Galileo,* ed. V. Coeltho, 45–63. Dordrecht: Kluwer, 1992.

Gingerich, O. "Kepler, Johannes," in *Dictionary of Scientific Biography,* ed. Charles Coulston Gillespie, vol. 7, 289–312. New York: Scribners, 1973.

Ginzburg, C. *The Enigma of Piero.* London: Verso, 2000.

James, J. *The Music of the Spheres.* New York: Copernicus, 1993.

Jardine, N. *The Birth of History and Philosophy of Science: Kepler's "A Defense of Tycho against Ursus" with Essays on Its Provenance and Significance.* Cambridge: Cambridge University Press, 1984.

Kepler, J. *The Harmony of the World.* Philadelphia: American Philosophical Society, 1997.

Kepler, J. *Mysterium Cosmographicum.* New York: Abaris Books, 1981.

Leonardo Da Vinci. NY: Artabras/Reynal and Company, 1938.

MacKinnon, N. "The Portrait of Fra Luca Pacioli," *Mathematical Gazette,* 77 (1993): 130–219.

Martens, R. *Kepler's Philosophy and the New Astronomy.* Princeton, NJ: Princeton University Press, 2000.

O'Connor, J. J., and Robertson, E. F. *www-history.mcs.st-andrews.ac.uk/history/Mathematicians/Pacioli.html/Durer.html.*

Pacioli, L. *Divine Proportion.* Paris: Librairie du Compagnonnage, 1988.

Pauli, W. "The Influence of Archetypal Ideas on the Scientific Theories of Kepler," in *The Interpretation of Nature and the Psyche,* 147–240. New York: Parthenon, 1955.

Stephenson, B. *The Music of the Spheres: Kepler's Harmonic Astronomy.* Princeton, NJ: Princeton University Press, 1994.

Strieder, P. *Albrecht Dürer.* New York: Abaris Books, 1982.

Taylor, R. E. *No Royal Road.* Chapel Hill: University of North Carolina Press, 1942.

Voelkel, J. R. *Johannes Kepler.* Oxford: Oxford University Press, 1999.

Westman, R. A. "The Astronomer's Role in the Sixteenth Century: A Preliminary Survey," *History of Science,* 18 (1980): 105–147.

7: PAINTERS AND POETS HAVE EQUAL LICENSE

Altschuler, E. L. *Bachanalia.* Boston: Little, Brown and Company, 1994.

d'Arcais, F. F. *Giotto.* New York: Abbeville Press Publishers, 1995.

Bellosi, L. *Cimabue.* New York: Abbeville Press Publishers, 1998.

Bergamini, D. *Mathematics.* New York: Time Incorporated, 1963.

Bois, Y.-A., Joosten, J., Rudenstine, A. Z., and Janssen, H. *Piet Mondrian.* Boston: Little, Brown and Company, 1995.

Boring, E. G. *A History of Experimental Psychology.* New York: Appleton-Century-Crofts, 1957.

Bouleau, C. *The Painter's Secret Geometry.* New York: Harcourt, Brace & World, 1963.

Curchin, L., and Fischler, R. "Hero of Alexandria's Numerical Treatment of Division in Extreme and Mean Ratio and Its Implications," *Phoenix,* 35 (1981): 129–133.

Curtis, W. J. R. *Le Corbusier: Ideas and Forms.* Oxford: Phaidon, 1986.

Duckworth, G. E. *Structural Patterns and Proportions in Vergil's Aeneid.* Ann Arbor: University of Michigan Press, 1962.

Emmer, M. *The Visual Mind.* Cambridge, MA: MIT Press, 1993.

Fancher, R. E. *Pioneers of Psychology.* New York: W. W. Norton & Company, 1990.

Fechner, G. T. *Vorschule der Aesthetik.* Leipzig: Breitkopf & Härtel, 1876.

Fischler, R. "How to Find the 'Golden Number' Without Really Trying," *The Fibonacci Quarterly,* 19 (1981): 406–410.

Fischler, R. "On the Application of the Golden Ratio in the Visual Arts," *Leonardo,* 14 (1981): 31–32.

Fischler, R. "The Early Relationship of Le Corbusier to the Golden Number," *Environment and Planning B,* 6 (1979): 95–103.

Godkewitsch, M. "The Golden Section: An Artifact of Stimulus Range and Measure of Preference," *American Journal of Psychology,* 87 (1974): 269–277.

Hambidge, J. *The Elements of Dynamic Symmetry.* New York: Dover Publications, 1967.

Herz-Fischler, R. "An Examination of Claims Concerning Seurat and the Golden Number," *Gazette des Beaux-Arts,* 125 (1983): 109–112.

Herz-Fischler, R. "Le Corbusier's 'regulating lines' for the villa at Garches (1927) and other early works," *Journal of the Society of Architectural Historians,* 43 (1984): 53–59.

Herz-Fischler, R. "Le Nombre d'or en France de 1896 à 1927," *Revue de l'Art,* 118 (1997): 9–16.

Hockney, D. *Secret Knowledge.* New York: Viking Studio, 2001.

Howat, R. *Debussy in Proportion.* Cambridge: Cambridge University Press, 1983.

Kepes, G. *Module, Proportion, Symmetry, Rhythm.* New York: George Braziller, 1966.

Larson, P. "The Golden Section in the Earliest Notated Western Music," *The Fibonacci Quarterly,* 16 (1978): 513–515.

Le Courbusier. *Modulor I and II.* Cambridge, MA: Harvard University Press, 1980.

Lendvai, E. *Béla Bartók: An Analysis of His Music.* London: Kahn & Averill, 1971.

Lowman, E. A. "Some Striking Proportions in the Music of Bela Bartók," *The Fibonacci Quarterly,* 9 (1971): 527–537.

Marevna. *Life with the Painters of La Ruche.* New York: Macmillan Publishing Co., 1974.

McManus, I. C. "The Aesthetics of Simple Figures," *British Journal of Psychology,* 71 (1980): 505–524.

Nims, J. F. *Western Wind.* New York: McGraw-Hill, 1992.

Nuland, S. B. *Leonardo da Vinci.* New York: Viking, 2000.

Osborne, H. ed. *The Oxford Companion to Art.* Oxford: Oxford University Press, 1970.

Putz, J. F. "The Golden Section and the Piano Sonatas of Mozart," *Mathematics Magazine,* 68 (1995): 275–282.

Sadie, S. *The New Grove Dictionary of Music and Musicians.* New York: Grove, 2001.

Schiffman, H. R., and Bobka, D. J. "Preference in Linear Partitioning: The Golden Section Reexamined," *Perception & Psychophysics,* 24 (1978): 102–103.

Schillinger, F. *Joseph Schillinger.* New York: Da Capo Press, 1976.

Schillinger, J. *The Mathematical Basis of the Arts.* New York: Philosophical Library, 1948.

Schwarz, L. "The Art Historian's Computer," *Scientific American* (April 1995): 106–111.

Somfai, L. *Béla Bartók: Compositions, Concepts, and Autograph Sources.* Berkeley: University of California Press, 1996.

Svensson, L. T. "Note on the Golden Section," *Scandinavian Journal of Psychology,* 18 (1977): 79–80.

Tatlow, R., and Griffiths, P. "Numbers and Music," in *The New Grove Dictionary of Music and Musicians,* 18 (2001): 231–236.

Watson, R. I. *The Great Psychologists.* Philadelphia: J. B. Lippincott Company, 1978.

White, M. *Leonardo the First Scientist.* London: Little, Brown and Company, 2000.

Woodworth, R. S., and Schlosberg, H. *Experimental Psychology.* New York: Holt, Rinehart and Winston, 1965.

Zusne, L. *Visual Perception of Form.* New York: Academic Press, 1970.

8: FROM THE TILES TO THE HEAVENS

Cohen, J., and Stewart, I. *The Collapse of Chaos. Discovering Simplicity in a Complex World.* New York: Penguin Books, 1995.

Fischer, R. *Financial Applications and Strategies for Traders.* New York: John Wiley & Sons, 1993.

Gardner, M. *Penrose Tiles to Trapdoor Ciphers.* New York: W. H. Freeman and Company, 1989.

Gleick, J. *Chaos.* New York: Penguin Books, 1987.

Lesmoir-Gordon, N., Rood, W., and Edney, R. *Introducing Fractal Geometry.* Cambridge: Icon Books, 2000.

Mandelbrot, B. B. *Fractal Geometry of Nature.* New York: W. H. Freeman and Company, 1988.

Mandelbrot, B. B. "A Multifractal Walk Down Wall Street," *Scientific American* (February 1999): 70–73.

Matthews, R. "The Power of One," *New Scientist,* July 10, 1999, 27–30.

Peitgen, H.-O., Jürgens, H., and Saupe, D. *Chaos and Fractals.* New York: Springer-Verlag, 1992.

Peterson, I. "Fibonacci at Random," *Science News,* 155 (1999): 376–377.

Peterson, I. *The Mathematical Tourist.* New York: W. H. Freeman and Company, 1988.

Peterson, I. "A Quasicrystal Construction Kit," *Science News,* 155 (1999): 60–61.

Prechter, R. R. Jr., and Frost, A. J. *Elliot Wave Principle.* Gainesville, GA: New Classics Library, 1978.

Schroeder, M. *Fractals, Chaos, Power Laws.* New York: W. H. Freeman and Company, 1991.

Steinhardt, P. J., Jeong, H.-C., Saitoh, K., Tanaka, M., Abe, E., and Tsai, A. P. "Experimental Verification of the Quasi-Unit-Cell Model of Quasicrystal Structure," *Nature,* 396 (1998): 55–57.

Stewart, I. *Does God Play Dice?* London: Penguin Books, 1997.

Walser, H. *The Golden Section.* Washington, DC: The Mathematical Association of America, 2001.

9: IS GOD A MATHEMATICIAN?

Baierlein, R. *Newton to Einstein: The Trail of Light.* Cambridge: Cambridge University Press, 1992.

Barrow, J. D. *Impossibility.* Oxford: Oxford University Press, 1998.

Chandrasekhar, S. *Truth and Beauty.* Chicago: University of Chicago Press, 1987.

Chown, M. "Principia Mathematica III," *New Scientist,* August 25, 2001, 44–47.

Davis, P. J., and Hersh, R. *The Mathematical Experience.* Boston: Houghton Mifflin Company, 1998.

Dehaene, S. *The Number Sense.* Oxford: Oxford University Press, 1997.

Deutsch, D. *The Fabric of Reality.* New York: Penguin Books, 1997.

Hersh, R. *What Is Mathematics, Really?* New York: Oxford University Press, 1997.

Hill, T. P. "The First Digit Phenomenon," *American Scientist,* 86 (1998): 358–363.

Kleene, S. C. "Foundations of Mathematics," Chicago: *Encyclopaedia Britannica* (1971), 1097–1103.

Lakatos, I. *Mathematics, Science and Epistemology.* Cambridge: Cambridge University Press, 1978.

Lakoff, G., and Núñez, R. *Where Mathematics Comes From.* New York: Basic Books, 2000.

Livio, M. *The Accelerating Universe.* New York: John Wiley & Sons, 2000.

Maor, E. *To Infinity and Beyond: A Cultural History of the Infinite.* Princeton, NJ: Princeton University Press, 1987.

Matthews, R. "The Power of One," *New Scientist,* July 10, 1999, 26–30.

Penrose, R. *The Emperor's New Mind.* Oxford: Oxford University Press, 1989.

Penrose, R. *Shadows of the Mind.* Oxford: Oxford University Press, 1994.

Pickover, C. A. *The Loom of God.* Cambridge, MA: Perseus Books, 1997.

Popper, K. R., and Eccles, J. C. *The Self and Its Brain.* New York: Springer International, 1977.

Raimi, R. "The Peculiar Distribution of the First Digit," *Scientific American* (December 1969): 109–119.

Raskin, J. www.jefraskin.com/forjef2/jefweb-compiled/unpublished/effectiveness_mathematics/

Robinson, A. "From a Formalist's Point of View," *Dialectica,* 23 (1969): 45–49.

Russell, B. *A History of Western Philosophy.* New York: Simon and Schuster, 1945.

Russell, B. *Human Knowledge, Its Scope and Its Limits.* New York: Simon and Schuster, 1948.

Weisstein, E. matworld.wolfram.com/BenfordsLaw.html.

Wolfram, S. *A New Kind of Science.* Champaign, IL: Wolfram Media, 2002.

INDEX

CREDITS

The author and publisher gratefully acknowledge permission to reprint the following copyrighted material:

ARTWORK:

Figs.1, 2, 3, 7, 9, 10, 11, 12, 14a, 14b, 18, 20a, 20b, 20c, 20d, 20e, 21, 24, 25a, 25b, 26, 27, 29, 30, 33a, 33b, 35, 37, 40, 41, 42, 44a, 44b, 49, 57a, 57b, 58, 61, 62, 63, 64, 86, 89, 91, 97a, 97b, 97c, 101a, 101b, 102a, 102b, 103a, 103b, 105, 106a, 106b, 107, 112, 114, 123, 124, and the diagrams in Appendix 2, Appendix 3, and Appendix 4 by Jeffrey L. Ward

Fig. 4: The Bailey-Matthews Shell Museum

Fig. 5: Chester Dale Collection, Photograph © 2002 Board of Trustees, National Gallery of Art, Washington, D.C. © 2002 Salvador Dali, Gala-Salvador Dali Foundation/Artists Rights Society (ARS), New York

Fig. 6: Reprinted with permission from John D. Barrow, *Pi In the Sky* (Oxford: Oxford University Press, 1992).

Fig. 13: © Copyright The British Museum, London.

Fig. 17: Hirmer Fotoarchiv

Fig. 19: Reprinted with permission from Robert Dixon, *Mathographics* (Mineola: Dover Publications, 1987).

Figs. 22 & 23, bottom: Reprinted with permission from H. E. Huntley, *The Divine Proportion* (Mineola: Dover Publications, 1970).

Fig. 23, top: Alison Frantz Photographic Collection, American School of Classical Studies at Athens

Fig. 28: Reprinted with permission from Trudi Hammel Garland, *Fascinating Fibonaccis Mystery and Magic in Numbers* © 1987 by Dale Seymour Publications, an imprint of Pearson Learning, a division of Pearson Education, Inc.

Figs. 31–32: Reprinted with permission from Trudi Hammel Garland, *Fascinating Fibonaccis Mystery and Magic in Numbers* © 1987 by Dale Seymour Publications, an imprint of Pearson Learning, a division of Pearson Education, Inc.

Fig. 34: Reprinted with permission from J. Brandmüller, "Five fold symmetry in mathematics, physics, chemistry, biology and beyond," in I. Hargitta, ed. *Five Fold Symmetry* (Singapore: World Scientific, 1992).

Fig. 36: Reprinted with permission from N. Rivier et al., *J. Physique,* 45, 49 (1984).

Fig. 38: The Royal Collection © 2002, Her Majesty Queen Elizabeth II

Fig. 39: Reprinted with permission from Edward B. Edwards, *Pattern and Design with Dynamic Symmetry* (Mineola: Dover Publications, 1967).

Fig. 43: Credit NASA and the Hubble Heritage Team.

Figs. 46, 45, 47, 50: Alinari/Art Resource, NY

Fig. 47: Perspective lines, reprinted with permission from Laura Geatti, Michelle Emmer Editor, *The Visual Mind: Art and Mathematics* (Cambridge: the MIT Press, 1993).

Fig. 52: Property of the Ambrosian Library. All rights reserved. Reproduction is forbidden.

Fig. 53: Scala/Art Resource, NY

Figs. 55, 56: The Metropolitan Museum of Art, Dick Fund, 1943

Fig. 57: Reprinted with permission from David Wells, *The Penguin Book of Curious and Interesting Mathematics* (London: The Penguin Group, 1997), copyright © David Wells, 1997.

Figs. 68–69: Kindly provided by the Institute for Astronomy, University of Vienna.

Figs. 70, 71, 72: Alinari/Art Resource, NY

Fig. 72: National Gallery, London

Fig. 73: Alinari/Art Resource, NY

Fig. 75: Scala/Art Resource, NY

Fig. 76: The Metropolitan Museum of Art, Bequest of Stephen C. Clark, 1960. (61.101.17)

Fig. 77: Philadelphia Museum of Art: The A. E. Gallatin Collection, 1952. © 2002 Artists Rights Society (ARS), New York/ADAGP, Paris

Fig. 78: Private Collection, Rome. © 2002 Artists Rights Society (ARS), New York/ADAGP, Paris

Fig. 79: © 2002 Artists Rights Society (ARS), New York/ADAGP, Paris/FLC

Figs. 80, 81: © 2002 Artists Rights Society (ARS), New York/ADAGP, Paris/FLC

Fig. 82: Private Collection. From "Module Proportion, Symmetry, Rhythm" by Gyorgy Kepes, George Braziller. © 2002 Artists Rights Society (ARS), New York/DACS, London

Fig. 83: The Museum of Modern Art/Licensed by Scala/Art Resource, NY. © 2002 Mondrian/Holtzman Trust, c/o Beeldrecht/Artists Rights Society (ARS), New York

Fig. 84: Reprinted with permission from G. Markowsky, *The College Mathematics Journal,* 23, 2 (1992).

Fig. 85: Reprinted with permission from Denis Arnold, ed., *The New Oxford Companion to Music,* Vol. 2 (Oxford: Oxford University Press, 1984).

Figs. 87, 88: Reprinted with permission from Ernö Lendvai, *Béla Bartók, An Analysis of His Music* (London: Kahn & Averill, 1971).

1987 by Dale Seymour Publications, an imprint of Pearson Learning, a division of Pearson Education, Inc.

ADDITIONAL CREDITS:

Fig. 8: Reprinted from Robert Lawlor, *Sacred Geometry* (London: Thames and Hudson, 1982).

Figs. 15–16: Reprinted from Robert Lawlor, *Sacred Geometry* (London: Thames and Hudson, 1982).

Page 195: Poem by Katherine O'Brien: Reprinted from Robert L. Weber, *Science with a Smile* (Bristol: Institute of Physics Publishing, 1992). All best effort was made to track down the source holder.

Page 195: Poem by J. A. Lindon: Reprinted from Martin Gardner, *Mathematical Circus,* (New York: Alfred A. Knopf, 1979). All best effort was made to track down the source holder.

Good faith efforts have been made to contact the copyright holders of the art in this book, but in a few instances the author has been unable to locate them. Such copyright holders should contact Broadway Books, a division of Random House, Inc. 1540 Broadway, New York, NY 10036.